1.20

150774902 1

AF744677

UNIVERSITY OF BRISTOL
...ING EDUCATION LIBRARY

GEOLOGY
FOR SCHOOLS

GEOLOGY FOR SCHOOLS

THIRD EDITION

J. T. GREENSMITH

B.Sc., Ph.D.

Senior Lecturer in Geology, Queen Mary College, University of London

INTERTEXT BOOKS

Published by
International Textbook Company Limited
24 Market Square, Aylesbury, Bucks

© J. T. Greensmith 1973

All rights reserved. No part of this publication may be reproduced, stored in a retrieval system, or transmitted, in any form or by any means, electronic, mechanical, photocopying, recording or otherwise, without the prior permission of the copyright owner.

First published 1964
Second edition 1967
Reprinted 1970
This edition 1973

ISBN 0 7002 0206 4

WITHDRAWN
University of Bristol

Reproduced photolitho in Great Britain by
J. W. Arrowsmith Ltd., Winterstoke Road, Bristol BS3 2NT

PREFACE TO THE FIRST EDITION

THE main branches of geology are introduced in this book in a fashion which it is hoped will prove suitable for systematic teaching at the Ordinary Level of the General Certificate of Education. The aim has been to make each chapter sufficiently comprehensive yet maintain an overall balance throughout the book appropriate to the subjects being reviewed. It was thought fitting that modern ideas should be incorporated as far as is possibly commensurate with the basic purpose of the book. The greatest problem has been that of condensation and hence generalization which means that many important facts have had to be glossed over or even omitted. For any lack of clarity or loss in accuracy created by these twin processes the author apologizes, as he does for any other inexactitudes which may arise. The questions at the end of each chapter are set in the style of the Ordinary Level G.C.E. They are a guide as to standards and are not intended to cover entirely the subject matter of the chapter. Finally, a 25 miles to 1 in. geological map of the British Isles should be consulted as often as possible during the reading of most of the chapters.

The motive power and inspiration in the creation of this book can be ascribed to my wife, who has assisted and encouraged me at all stages. Her experience as a school teacher and trained geologist has been invaluable not only in correcting and clarifying the text but also in selecting and drawing many of the diagrams. Miss S. Downing typed out all the original drafts. Professor J. F. Kirkaldy was also helpful on many of the technical problems arising from writing a book.

CONTENTS

CONTENTS

LIST OF PLATES

LIST OF FIGURES

Chapter 1

INTRODUCTION

UNTIL recently geology used to be defined as the study of the history of the earth since its inception more than 5,000 million years ago. In this age of space travel, however, the old definition requires modification because there is little doubt that among the first travellers to the Moon and more distant planets will be geologists. Geology concerns itself with natural processes and effects which are visible today at the surface or can be inferred from the depths and which were equally visible and deducible millions of years ago, though obviously not to man. In doing so it relies on a fundamental law of nature which states that the external and internal processes affecting rocks, organisms and landscapes at present are no different from those in the past. This is the Principle of Uniformitarianism and it was first expounded by a British geologist, Charles Lyell, well over a hundred years ago.

In attempting to unravel the past geology draws heavily on all the other major sciences. Chemistry and physics help in identifying and classifying minerals (**mineralogy**) and rocks (**petrology**). They are equally important in working out the processes of weathering, erosion, transportation and deposition (**physical geology**). The electrical, gravitational and elastic properties of rocks at outcrop and at depth are determined by geophysical means and from this data rock structures can be deduced (**structural geology**). Biology is very useful too in the classification of animals and plants and in indicating evolutionary processes and patterns of organic distribution applicable to the past (**palaeontology**). Statistical mathematical techniques are an important tool in this and other geological work.

A large body of information about rocks is built up, therefore, using the techniques of related sciences. However, the final interpretation of the history of the rocks rests squarely on the shoulders of the geologist. He has to decide on the relative ages of the rocks, their manner of formation and their patterns of distribution at any one time (**stratigraphy**). This is not all, because his labours are occasionally rewarded by the discovery of materials of economic value such as uranium-bearing veins, coal and ironstone seams, rock-salt deposits or oil pools. In fact the main function of the commercial geologist is to find these materials in such quantities as to make them worthy of exploitation. There are still large tracts of the earth's crust which have not been adequately explored either from a commercial or from a more strictly academic point of view. There is ample scope for geological research and much work is in hand at present. Geological knowledge is increasing constantly and with it new ideas. The findings of the major nineteenth-century geologists have been subject to many modifications because of this new information but it is much to their credit that most of the earliest generalizations in all branches of geology have the same validity now as they had in the past.

Chapter 2

MINERALS

THE mineral wealth of a country is one of the important factors which can affect its economic growth. Deposits of coal, iron ore, salt, gold, silver and oil immediately spring to mind. To the geologist, however, only some of these natural products are minerals in the strict sense. This is because a mineral is usually defined as a naturally-occurring substance with a definite composition and constant physical properties. As deposits such as coal, iron ore and oil have a wide range in composition and far from constant physical properties they do not fall within this definition. On the other hand it has to be realized that minerals do vary slightly in their properties but the variation is strictly limited by their internal arrangements of atoms. This nearly constant composition of any given mineral can be expressed by means of a chemical formula which can be either simple or complex depending on the number of elements present.

Habit

Perhaps the most important feature of minerals, certainly for simple identification, is their external shape. This is especially so if the mineral grains have smooth flat faces as do many in veins and natural cavities. These well-formed minerals with smooth surfaces are called **crystals,** and by working out the number and different types of face and measuring the angles between them it is often possible to classify and even sometimes name the minerals.

The outer shape of crystals may be thought to be entirely related to their fundamental internal structure—the arrangement of atoms—but in reality the development of shape is dependent mainly on the freedom with which faces are allowed to grow. A given mineral may have many shapes therefore and these are referred to as its **crystal habits** (Fig. 1). If, for instance, growth occurs freely in all directions and the faces are all more or less similar in size, then an **equant** habit forms, e.g. garnet, fluorspar, galena. If, in contrast, a pair of opposite faces enlarge at a faster rate than the others then a **tabular** habit tends to form as in barytes and feldspar. In some cases this tabular tendency is carried to extremes, and a very thin flat crystal is developed with a **lamellar** or **foliated** habit. Micas are good examples of this. An ordinary unsharpened multi-sided pencil can be used to illustrate yet another common habit. This is the **columnar** habit in which three or more of the side faces paralleling the long axis of the pencil are elongated and enlarged relative to the top and bottom faces. Quartz often takes on this appearance. In extreme cases this habit is called **acicular** (needle-shaped), e.g. fibrous gypsum.

These then are terms which can be used in describing minerals with good flat surfaces. How are minerals described when their

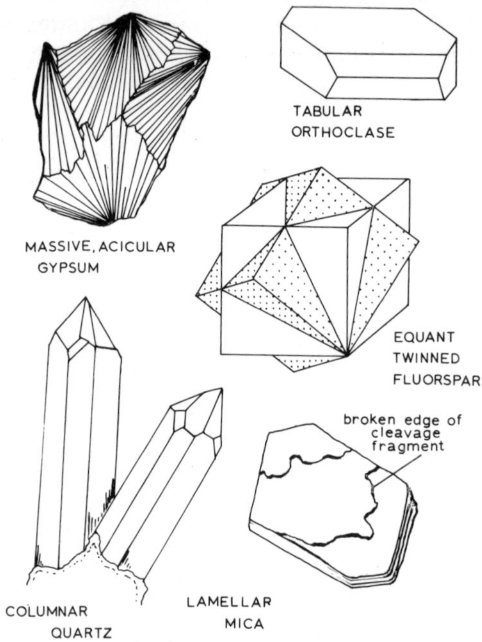

FIG. 1. Common crystal habits

faces are irregular, obscure or even absent? Many minerals commonly exhibit irregular external shapes, so additional terms are clearly needed. There is, fortunately, a wide range to select from and, in practice, the real problem comes in selecting the most accurate term (or terms) from this range. An example of this type of problem can arise with the green copper carbonate malachite which can have a **mamillated** habit consisting of large mutually-interfering spheroidal surfaces or a **botryoidal** habit (Fig. 2) which resembles a bunch of grapes. On inspection of the mineral the term which might be selected could easily be either, depending on the impression gained by the viewer. In other words many of these terms are subjective. A further matter which can cause confusion is that one mineral specimen can exhibit more than one habit. Iron pyrites often occurs as irregular nodules (**nodular** or **concretionary** habit) and when the nodules are broken across a **radiating** (Fig. 2) habit may be exposed to view. This is an arrangement of the mineral fibres in which they appear to diverge from a single central point.

Native copper and pyrolusite (manganese oxide) sometimes have a curious tree-like appearance which is called **arborescent** or **dendritic** (Fig. 2); it is not a common habit of many minerals. Likewise a **reniform** or kidney-shaped habit seems to be restricted and is usually associated with minerals deposited colloidally from

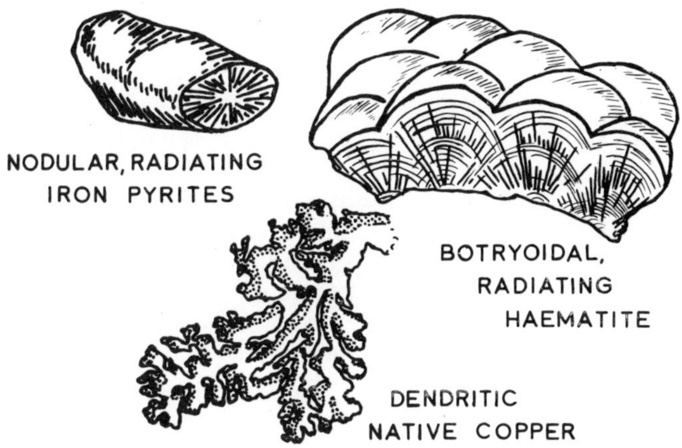

FIG. 2. Common mineral habits

solution, e.g. haematite. Botryoidal and mamillary masses also originate in that fashion.

Perhaps the most important point about all the habits mentioned above, remembering that they are only a selection, is that they are used in referring to one specific mineral, not a mixture of minerals. The mineral may be crystalline with either a flat-faced or irregular outer form, or it may be non-crystalline (amorphous) without flat faces. Opal and limonite are typical amorphous minerals.

Quite frequently several crystalline grains or amorphous aggregates of one mineral may be so closely intergrown that no distinctive habit can be seen. In these cases the general term **massive** is used. Practically all the common minerals can adopt a massive habit.

Twinning

A useful diagnostic property of some minerals is referred to as **twinning.** It can be well seen in hand specimens in the feldspar group. The fluorspar illustrated in Fig. 1 shows how it can be detected sometimes. In the diagram it can be seen that instead of one equant grain there appears to be two which have almost completely intergrown giving a marked modification of each of the six flat faces of the cube. This type of twinning is called **interpenetration twinning** and as it only concerns what seem to be two parts of the same mineral it is also **simple twinning.** There are many other kinds of twinning in minerals but they are the subject for more advanced studies.

Colour

The colour of minerals is something which, rather surprisingly, can be dismissed quickly. This is simply because colour is often caused by impurities within a mineral. The red colour of some feldspars is commonly due to very small amounts of haematite; likewise the colours of rose and amethystine quartz are probably due to traces of manganese and titanium. The impurities in fluorspar can give a range of colours of which purple, blue, yellow and white are merely a selection. On the other hand the colour of some minerals can be quite diagnostic as in the cases of yellow native sulphur, green malachite and blue azurite. However, this property has to be used with caution, as a study of hand specimens will show.

Streak

More diagnostic than the colour of a mineral is the colour of the fine powder produced when it is ground down. This can be done on a small, though adequate, scale by rubbing the mineral on a piece of unglazed porcelain called a streak plate. The results are interesting as the streak colour may often differ from the hand specimen colour. Haematite can have a steel-grey to black colour but gives a distinctive red streak. Another common iron mineral, limonite, which can have a blackish-brown appearance gives a brownish-yellow to orange-yellow streak.

Lustre and Tarnish

The reflection of light from a mineral is called **lustre.** It is a natural property of minerals yet it can be modified in many ways as, for instance, in the gemstone industry where polishing and cutting can enhance the lustre. Diamonds in their natural state have a **greasy** lustre due to roughness of their surface; when polished they have an **adamantine** (or **brilliant**) lustre. Rock salt (halite) in its fresh state has a **vitreous** (or **glassy**) lustre but tarnishing on exposure gives a greasy or **waxy** lustre. Other commonly used terms are **resinous, pearly, dull** (or **earthy**), **submetallic** and **metallic.**

Hardness

Tarnish not only affects the lustre of minerals but also the apparent hardness; it usually makes a mineral seem softer than it actually is. Therefore when testing for hardness a clean fresh surface must be used. Hardness can be defined as a resistance to scratching and is normally recorded in terms of the following scale of increasing hardness proposed by an Austrian mineralogist, Mohs, in 1822:

1 Talc	6 Orthoclase
2 Gypsum	7 Quartz
3 Calcite	8 Topaz
4 Fluorspar	9 Corundum
5 Apatite	10 Diamond

Each of these relatively common minerals can scratch the ones lower down in the scale and can be scratched by those above. If a mineral being described could be scratched by quartz and not by apatite its hardness would fall between 5 and 7. A useful point to remember is that in the absence of a set of Mohs minerals other common objects can be used. The average hardness of a finger nail is $2\frac{1}{2}$ and a penknife usually has a hardness of $5\frac{1}{2}$.

Fracture and Cleavage

When most mineral specimens are examined closely, there can be seen irregular surfaces which are clearly not the faces of crystals. These rough surfaces are also found in amorphous grains and are **fractures.** A common type of fracture, particularly in amorphous minerals and crystalline quartz, is **conchoidal,** consisting of a concentric series of curved surfaces. Other types are **splintery, even, uneven** and **hackly** (sharp and jagged).

Cleavage is a more fundamental property and originates in internal weaknesses in the structure of crystalline minerals. Amorphous minerals do not cleave. The presence of cleavage is usually indicated by smooth, flat, lustrous surfaces which become visible when a crystal is split open. These cleavage surfaces must not be confused with the smooth outer faces of crystals.

Quartz, for instance, often has flat outer faces but it has **poor** cleavage. The micas have one **perfect** cleavage direction but other minerals such as rock salt and galena have three, all at right angles to each other. There are also minerals which have multiple cleavage directions of unequal quality. The feldspars have one perfect and one **good** cleavage. In contrast tourmaline has poor or **imperfect** cleavage with very little ease of splitting along any definite plane.

Other properties

There are many other properties of great importance, but they can only be diagnosed successfully in specially-equipped laboratories. The most important factor in the accurate determination of these properties is purity in the specimen.

Chemical composition is determined by various kinds of analysis some of which are time-consuming. Modern techniques are tending to speed analysis and many can be done very quickly.

The crystal system or grouping to which a mineral belongs is assessed by means of X-rays or more laboriously and cheaply by means of a petrographic microscope.

The density or specific gravity can be determined in many ways. One of the simplest is the suspension method, which consists of immersing the mineral grain in liquids of known density until it neither floats to the surface nor sinks to the bottom. The density of the mineral is the same then as the density of the immersion liquid. Few of the common minerals are sufficiently low or high in density to enable a positive identification to be made in hand specimens. One of the exceptions is barytes (density 4·48) which feels relatively heavy even as small grains.

Finally, there are two simple tests which are sometimes useful in identification. The first is the reaction with dilute hydrochloric acid. Calcite effervesces strongly, and iron pyrites generates hydrogen sulphide gas. The second is taste, and is particularly appropriate for detecting rock salt.

SOME COMMON MINERALS

Minerals are grouped in many complicated ways, mainly on the basis of their composition and internal structure. A very simple and convenient grouping is adopted here with a two-fold division into non-silicate and silicate minerals. As these minerals should be examined and described by looking at actual specimens, only the more important properties (including mode of occurrence) are mentioned.

Non-silicate Minerals

Sphalerite (or Zinc Blende), ZnS, four-sided crystals common, black or brown, perfect cleavage, hardness 3½ to 4, resinous to adamantine lustre. Hydrothermal veins and masses. Most important source of zinc.

Galena, PbS, cubic crystals, lead-grey, perfect cubic cleavage, metallic lustre. Hydrothermal veins and masses. Most important source of lead.

Iron pyrites, FeS_2, cubic crystals with striations, brass-yellow, hardness 6 to 6½. Hydrothermal veins; nodules in sediments. Most common sulphide mineral.

Copper pyrites, $CuFeS_2$, commonly massive, bronze-yellow, tarnishes easily, hardness 3½ to 4. Hydrothermal veins and masses. Most common copper mineral.

Native copper, Cu, arborescent, light pink tarnishing to copper-red to brown, hackly fracture, malleable. Hydrothermal veins; cavity infillings in basic igneous rocks.

Malachite, $CuCO_3.Cu(OH)_2$, massive, botryoidal, bright green. Secondary mineral in veins in limestone areas.

Azurite, $2CuCO_3.Cu(OH)_2$, massive, azure to dark blue, earthy lustre. Secondary mineral associated with malachite.

Magnetite, Fe_3O_4, massive, black, black streak, magnetic. Widespread in igneous and metamorphic rocks; veins.

Haematite, Fe_2O_3, massive, reniform, steel-grey to bright red, red to red-brown streak, metallic to submetallic lustre. Hydrothermal veins; sediments. Most important iron mineral.

Limonite (or Goethite), FeO(OH), massive, botryoidal, stalactitic with internal fibrous, radiating or concentric structure, blackish-brown to yellow, brownish-yellow to orange-yellow streak. A weathering product of all other iron minerals. Common in all rocks.

Cassiterite, SnO_2, massive, prismatic, yellow or red-brown to brown-black, hardness 6 to 7, heavy (density 6·99). Hydrothermal veins; river gravels. Most important source of tin.

Graphite, C, black colour and streak, greasy lustre, hardness 1 to 2. Layers and patches in metamorphic rocks; hydrothermal veins.

Sulphur, S, massive, prismatic, yellow, hardness 1½ to 2½, brittle. Deposits in volcanic areas; beds in salt sequences.

Calcite, $CaCO_3$, hardness 3, perfect rhombic cleavage, vitreous lustre, effervesces freely in cold dilute hydrochloric acid. The main mineral of limestones.

Dolomite, $CaCO_3.MgCO_3$, rhombs with curved faces, hardness 3½ to 4, very slight effervescence with cold dilute hydrochloric acid. Common in limestones.

Siderite, $FeCO_3$, yellowish-brown, heavy (density 3·96), alters readily to brown limonite. Common in sediments; hydrothermal veins.

Rock salt (or Halite), NaCl, massive, cubes, cubic cleavage, hardness 2½, salty taste. Sediments.

Gypsum, $CaSO_4.2H_2O$, massive (alabaster), tabular (selenite), fibrous (satin spar), one perfect and two poor cleavages, hardness 2. Sediments.

Anhydrite, $CaSO_4$, massive, hardness 3½. Sediments.

Barytes, $BaSO_4$, tabular, prismatic, heavy (density 4·48). Hydrothermal veins; gangue.

Fluorspar (or Fluorite), CaF_2, cubes, perfect cleavage, hardness 4. Hydrothermal veins; gangue.

Silicates

Quartz, SiO_2, columnar, massive, no cleavage, hardness 7. Most rocks; veins; gangue.

Opal, $SiO_2.nH_2O$, massive, botryoidal, hardness 5½ to 6½. Hydrothermal veins and pockets; hot springs.

Chalcedony, a mixture of fine-grained quartz and opal, massive (chert, flint), banded (agate, onyx). Sediments.

Olivine, $(Mg,Fe)_2SiO_4$, massive, olive-green, vitreous lustre, alters readily to serpentine and limonite. Igneous rocks.

Augite, Ca,Mg,Fe,Al silicate, tabular, black, two good cleavages at 89°. Igneous rocks.

Hornblende, Na,Mg,Fe,Al silicate, tabular, black, two good cleavages at 56°. Igneous and metamorphic rocks.

Biotite (or Black Mica), K,Mg,Fe,Al silicate, lamellar, one perfect cleavage, hardness 2½. Igneous and metamorphic rocks.

Muscovite (or White Mica), K,Al silicate, properties similar to biotite. Most rocks.

Orthoclase, $KAlSi_3O_8$, prismatic, two good cleavages, simple twinning, hardness 6. Igneous and metamorphic rocks; sandstones.

Plagioclase, Na,Ca,Al silicate, properties similar to orthoclase but multiple twinning. All rocks.

Tourmaline, B,Al silicate, columnar, triangular cross-section, acicular, vertical striations. Veins; altered igneous rocks.

Garnet, Ca,Mg,Fe,Mn,Al silicate, equant, massive, hardness 7 to 7½, no cleavage. Metamorphic rocks; veins.

QUESTIONS

1 Define the term mineral. Select three non-silicate and three silicate minerals and explain how you would distinguish between them.

2 Distinguish between ore minerals and gangue minerals by reference to a mineral vein containing galena, fluorspar, pyrites, sphalerite and barytes.

3 Describe five different minerals in which one or other of the following properties are diagnostic: (*a*) streak, (*b*) rhombic cleavage, (*c*) lamellar habit, (*d*) colour, (*e*) multiple twinning.

Chapter 3

ROCKS

THE systematic study and classification of rocks, or **petrology,** commenced towards the end of the 1700's but received its greatest impetus in the period 1850–1890 when it was realized that microscopes could reveal important details not noticeable in hand specimens. Not only could the microscope help to identify positively the actual mineral composition of rocks but it could also clarify the **textures** present. Textures are small-scale features of rocks; they include grain size, evenness of grain size and colour. All these could now be accurately described—a type of study called **petrography.** At the same time it became clear that the microscope was going to be of great assistance in field studies for working out the mode of origin or **petrogenesis** of rocks and today the instrument is regarded as an essential tool in all petrological work.

As a result of all this detailed field and laboratory work, the latter involving the accumulation of much chemical data, it is now certain that rocks can be classified into three main groups: **igneous, sedimentary** and **metamorphic.** Igneous rocks are those which have formed on the surface or within the earth's outer layers mainly from the consolidation of hot, sticky fluid material called **magma.** Magma is believed to be generated by melting of the lower layers of the earth's crust. Sedimentary rocks are formed from the physical and chemical disintegration of all other rocks followed by reconstitution and hence are derivative or secondary. Metamorphic rocks are created by the 'in place' alteration of pre-existing rocks of any type. For instance, metamorphic rocks themselves are often re-metamorphosed. The alteration brought about varies from slight to intense and is caused by a combination of high temperatures, high pressures and chemically-active fluids, any of which may be dominant.

IGNEOUS ROCKS

One of the important features of igneous activity is that it is commonly found in areas of crustal instability. At present the margins of the Pacific Ocean, the East and West Indies and the Mediterranean Sea area are four unstable zones, relatively narrow and elongated, where earth-tremors and vulcanicity are commonplace. Volcanic activity has been defined by Holmes (1944) as all the phenomena associated with the escape at the earth's surface of magmatic materials from the depths. It is a definition which clearly refers to the actual physical features at the earth's surface called volcanoes and to the hidden feeder-pipes or -lines in the upper layers of the crust through which the magma pours. Large-scale and deeper-seated igneous bodies such as batholiths (p. 19) appear to be excluded. On the other hand, many geologists think that volcanic activity (or vulcanicity) should include all igneous effects, whether surface or sub-surface.

Extrusive Rocks

Sometimes the eruptions of molten material create spectacular effects on land-surfaces and one of the best known is the volcano. Many volcanoes are very explosive in character but even so the explosions may be very spasmodic with long intervals, perhaps decades, of quiescence or dormancy. The more viscous (or sticky) the magma ascending from the depths, the more spasmodic and violent are the phases of activity as has been the case with Stromboli in Sicily and Vesuvius in Italy. Viscous magmas also build up steep-sided cones. Apart from igneous rocks cones commonly have a high proportion of **volcanic breccia** (Pl. 1) or **agglomerate** and **ash.** These rocks are **pyroclastic,** that is, fragmental in origin. They have been created by the physical break-up of magma during explosions and deposited adjacent to the **vent** after transport through the air. The coarser angular breccias are found nearest to the vent and sometimes even partially infill it, whereas the finer ashes have a wider distribution. The ash grains are often influenced in their distribution by the prevailing winds and the local topography of the adjacent countryside. When larger explosions occur the ash may be carried many hundreds of miles before being deposited. Ash after deposition and consolidation is called **tuff.**

Volcanoes often build up from the sea-bed and eventually may emerge above water-level to the accompaniment of great rumblings and explosions. Some of the larger explosions are probably due to the rapid generation of high-pressure steam when the hot magma comes into contact with cold sea-water.

In contrast the extrusion of magma may be relatively quiet both on land and in the sea. Very large flattish cones have been built up in this way, especially in the central parts of the Pacific Ocean. Mauna Loa in the Hawaiian Islands consists of at least 28,000 ft. of relatively quietly-deposited lavas and thin ash-beds, of which 14,000 ft. are above sea-level. Near to the summit is the focus of many of the lava-flows. It consists of a 1,000-ft. deep faulted and collapsed depression called a **caldera** which is about two miles wide and contains lakes of molten lava which occasionally rise in height and spill over the irregular edge of the depression. Sporadic explosions occur if the lava comes into contact with water. Apart from having several main feeder-vents

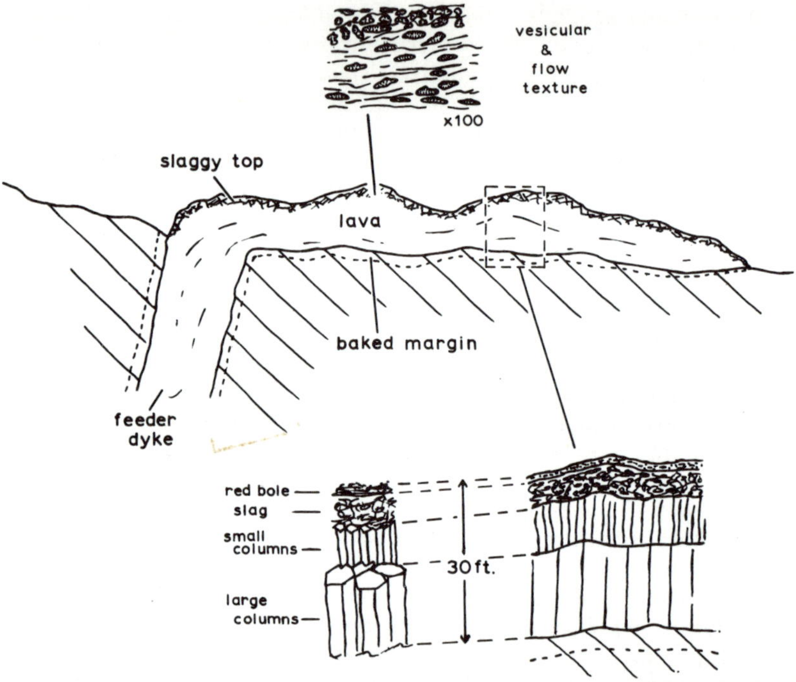

FIG. 3. Common structures and textures in lava-flows. In submarine lavas the slaggy top may be absent and there is no weathering to red bole. Columnar jointing is very poorly developed in some thin lava-flows

or **pipes** Mauna Loa is also interesting in that many of the lavas issue forth from a number of near-parallel north-east to south-west fissures down the sides of the cone. The volcano is quite complicated, therefore, in structure and origin.

Fissure extrusion is also commonplace in Iceland at present and occasionally **parasitic cones,** 50 to 500 ft. high, build up along the line of the fissures. It is possible that the Tertiary basalt lavas forming the Antrim Plateau and occurring in the Inner Hebrides area are of the fissure-fed type, the magma having been extruded from a major swarm of north-west to south-east trending feeder dykes. Alternatively, the lavas may have been erupted from large volcanoes of the Hawaiian type with foci in Skye, Rum, Mull, Arran and Ardnamurchan.

Whether erupted via volcanic vents or fissures, lavas are the best example of extrusive rocks. As they are extruded into air and water it is hardly surprising that they cool rapidly. The top layers on land-flows rapidly form a slag or scum which congeals to give a blocky or ropy surface. Simultaneously the gases present within the lava ascend towards the top, resulting in a cindery porous texture called **vesicular** (Fig. 3). The individual cavities or **vesicles** may be several inches long and have a tendency to be elongated in the direction of flow. Sometimes the flow creates almond-shaped cavities; these are **amygdales** and if abundant the rock texture is referred to as **amygdaloidal.** Vesicles and amygdales are commonly infilled with secondary precipitated minerals such as calcite and chalcedony.

Primary minerals as well as cavities often show alignment or **flow structure** in lavas. This is very well seen in hand specimens of rhyolites and trachytes but less so in andesites and basalts (Fig. 4). The flow structure actually observed is not always obviously due to minerals because lavas are fine-grained and it is often difficult to see the minerals even with a microscope. It is, in fact, an aggregate mineral effect. The smallness of grains in lavas results from fast cooling; occasionally the cooling is so rapid that no minerals form at all and a natural glass is produced. Obsidians and pitchstones are examples of natural glasses. Many interesting textures are found in these black rocks. In the last stages of consolidation a spheroidal series of fractures may develop; this is **perlitic** texture (Fig. 4). In other instances attempts at recrystallization give rise to **spherulitic** texture (Fig. 4), in which the individual spheres of semi-crystallized minerals are commonly a quarter of an inch across but sometimes reach a diameter of several feet. Natural glasses may be completely recrystallized or devitrified with the passage of time and because of this there are very few preserved in rock successions older than the Mesozoic.

In many lavas, especially andesites and basalts, the cooling of the magma seems to have taken place in two stages. The visible evidence for this is the presence of relatively large grains, **phenocrysts,** set in a much finer-grained ground-mass. It is probable that the phenocrysts represent a slow rate of cooling within the magma at some depth in the crust prior to the magma being squirted out on to the surface where cooling is much more rapid.

The resultant texture is **porphyritic** (Fig. 4). Non-extrusive rocks such as granite can also have this texture.

Perhaps the most spectacular cooling-contraction structure of lavas is **columnar jointing** (Fig. 3 and Pl. 2) and this is particularly well seen at the present day along the coast of Antrim. The joints are at right angles to the top and bottom cooling surfaces. In many cases these Tertiary basalts consist of three distinct layers. At the base are thick, vertical prismatic columns with a regular six-sided shape. They are up to two feet in diameter. In the middle is a layer of thinner, variable-sided columns and on top of that a slaggy, vesicular layer. The upper slaggy zone is of some further interest as it often shows evidence of subaerial weathering under subtropical climatic conditions. The evidence consists of the presence of a lateritic soil profile or 'red bole' rich in alumina.

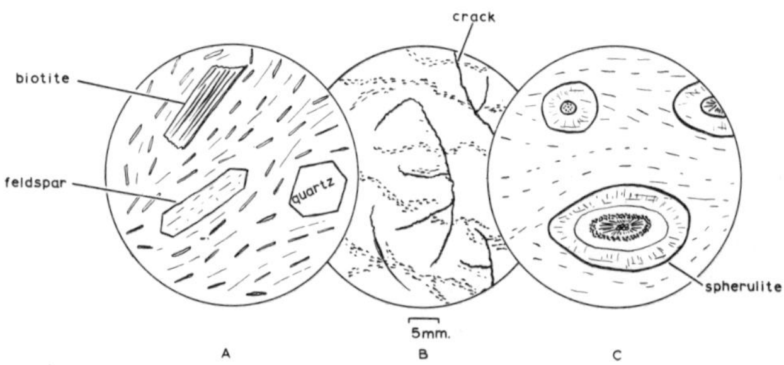

FIG. 4. Common textures in lavas. A: porphyritic with large phenocrysts set in a flow-aligned finer-grained matrix. As quartz is present this is an acid flow. B: perlitic. Spherical cracks cross-cutting primary flow texture. C: spherulitic. Secondary fibrous structures cross-cutting primary flow texture

The boles have been worked in the past as a source of aluminium. As they are softer and more readily weathered they create a marked scenic effect known as **stepped** or **trap topography.** A similar type of trap topography can be observed in the Campsie Fells and Clyde Plateau area of the Midland Valley of Scotland where the basalts are Carboniferous in age.

Minor Intrusive Rocks

In contrast to the extrusive form of lava-flows, their feeder channels from the depths are entirely **intrusive.** These feeders are **dykes** (Fig. 5) and they are normally near-vertical at the time of their emplacement and thus show a marked discordance to the surrounding rocks. It must not be thought, however, that all dykes feed lava-flows. In many cases the magma never reaches the surface at all.

The width of igneous dykes varies quite considerably but a reasonably average figure is 15 ft. Many persist over long distances. Some of the east–west trending Carboniferous dolerite dykes of Scotland can be traced more or less continuously for 30 to 40 miles. They tend to form narrow hump-backed ridges because they are harder than the adjacent sediments. The mid-Tertiary Cleveland Dyke extends from near the Yorkshire coast at Whitby westwards into the Carlisle Plain then, under another name, it swings north-westwards towards Arran, cutting through the Southern Uplands of Scotland. The distance is more than 150 miles.

A feature of many regions of intense volcanic activity, such as the western coast of Scotland during Tertiary times, is the presence of a large number, sometimes hundreds, of dykes all

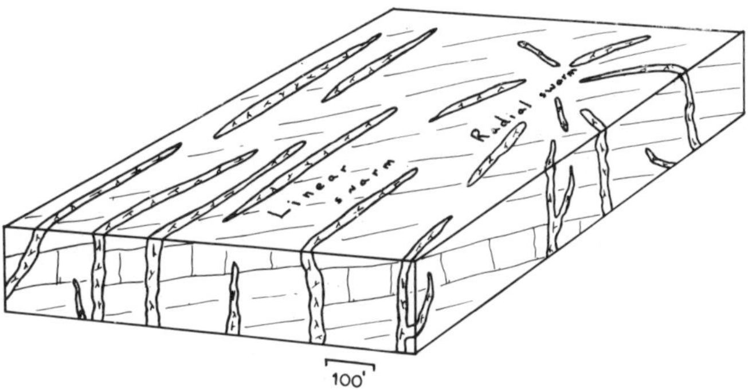

FIG. 5. Dykes. Note the cross-cutting relationships and the fact that some dykes never reach the surface during their emplacement

roughly elongated in one direction. The parallel arrangement is a **linear dyke swarm** (Fig. 5). **Radial dyke swarms** are much rarer (Fig. 5). The best-known British example is around the Cheviot granite where andesite dykes of Old Red Sandstone age radiate outwards like the spokes of a wheel.

With either kind of swarm it is probable that crustal warping generated by deep-seated earth-movements has created fissures into which the magma has flowed. The linear trend of the Scottish Tertiary dolerite swarm suggests that tension existed in the crust in an ENE–WSW direction and this produced a NNW–SSE series of fissures. The tension may have been caused by a warping of the continental shelf (p. 94) and on the Island of Arran it has been calculated that the crust was stretched 6,000 ft. at the time. Further to the north in Mull the crustal stretch was at least 2,500 ft.

If a dyke, especially a thicker one, is examined in the field it is sometimes possible to see a gradation in grain size from the central parts towards the margins. In the centre the size of the minerals may be almost 1 mm. whereas at the edges the individual grains may not be visible even with a hand lens. The variation is due to a chilling effect created by the relatively cold country rocks into which the hot magma has been injected. A dyke, therefore, has two **chilled margins.** A related effect to chilling is **baking** and every dyke has immediately adjacent to it on both sides baked zones of country rock. The width of these two baked zones depends on the heat and chemical composition of the magma, its rate of cooling and the nature of the country

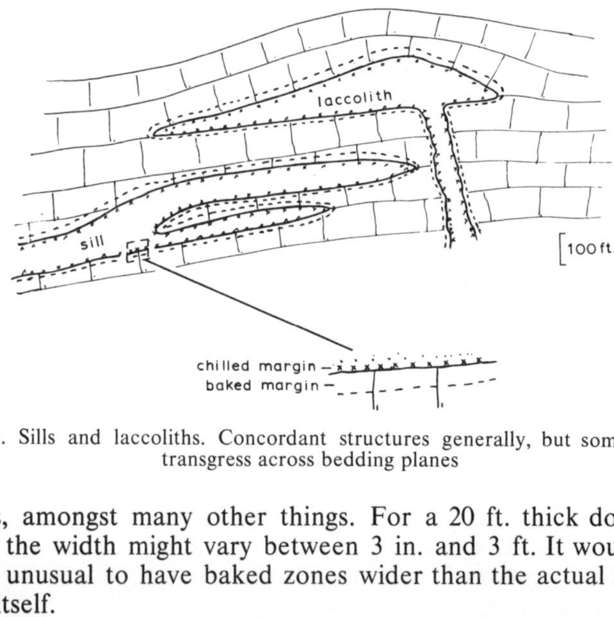

FIG. 6. Sills and laccoliths. Concordant structures generally, but sometimes transgress across bedding planes

rocks, amongst many other things. For a 20 ft. thick dolerite dyke the width might vary between 3 in. and 3 ft. It would be most unusual to have baked zones wider than the actual intrusion itself.

The presence of baked margins is useful in the field when deciding whether an igneous body is a lava-flow or a similarly concordant body called a **sill** (Fig. 6). The Hercynian Whin Sill of northern England is recognizable as a sill because it has a thin baked zone below and a thicker baked zone above. If it were a lava-flow only the lower zone would be present. The Whin Sill has an average thickness of 100 ft., but this is not unusual for Britain where some sills attain a thickness of 250 ft. In thicker sills columnar jointing is commonplace. Most major sills when traced over large distances split into several layers which may independently thin and thicken or even disappear entirely. The dolerite and gabbro Carboniferous sills of the Midland Valley are of this type.

A slightly modified form of sill is called a **laccolith** (Fig. 6). This can be compared for shape with either an inverted soup- or cereal-plate or a much-flattened hemisphere with its convex side upwards. The maximum diameter of laccoliths is normally five to six times the thickness. Some of the Ordovician laccoliths of central Wales are interesting in that they each have a vertical feeder dyke. The magma apparently squirted up a vertical fissure before spreading out along the bedding planes of the sedimentary rocks. One of the major problems about these and other laccoliths from further afield is whether the shape of the intrusion is related to the actual force of the intrusion or due to a natural gaping of the country rocks during simultaneous earth-movement and igneous activity.

Major Intrusive Rocks

Dykes, sills and laccoliths are relatively minor examples of intrusive igneous rocks despite the fact that they are widespread at high levels in the earth's crust. Much larger in volume and area are intrusive masses known as **lopoliths** and **batholiths** (Fig. 7).

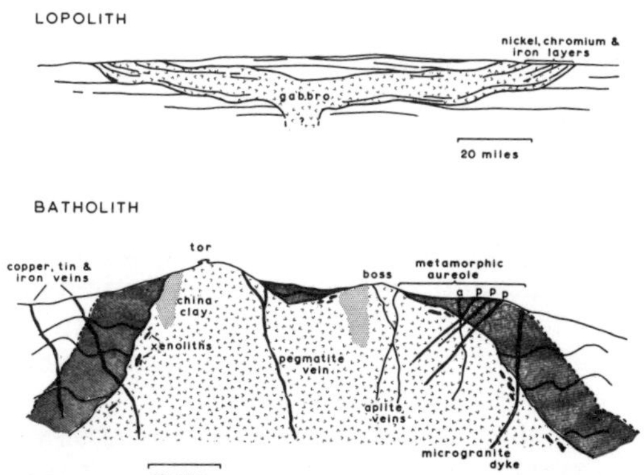

FIG. 7. A lopolith is a large-scale layered transgressive structure composed of acid to ultrabasic rocks though gabbros are dominant. Batholiths are very large-scale transgressive structures. Aplites (*a*) are very fine-grained rocks similar in composition to the parent granite and pegmatites (*p*). All the veins in the figure are too wide in relation to scale.

Lopoliths are basin-shaped masses and are exposed most frequently in shield areas of the world composed of Pre-Cambrian basement rocks. At the southern end of the Canadian Shield in the Lake Superior area the Duluth lopolith measures about 150 miles across, is nearly 10 miles thick and contains 50,000 cubic miles of rock. Not all this igneous rock is of one type, although coarse-grained basic rocks of the gabbro kind are dominant. The Bushveld lopolith in South Africa is of a similar size and internal complexity. One of the most characteristic features of all these masses is an internal lamination or layering which roughly parallels their top and bottom surfaces. The layers vary from less than one inch to several feet in thickness and are caused by differences in mineral composition. The light-coloured or **leucocratic** layers have a greater proportion of light-coloured **felsic** minerals such as feldspar; the dark or **melanocratic** layers have more dark **mafic** minerals such as hornblende, augite and olivine. At Duluth and elsewhere the dark layers are sometimes rich in minerals of economic value such as pyrites, nickel sulphides and chromium oxides.

Batholiths are even larger bodies of rock and are usually formed of granite. These more than any other igneous body are found in areas of the world where there has been abnormal crustal thickening and mountain building. The core of the great mountain ranges extending down the western coasts of North and South America is occupied by many granite batholiths some of which are 1,000 miles long by 100 miles wide. In the Central Highlands of Scotland a whole series of smaller batholithic masses extends from Peterhead and Aberdeen south-westwards into the Ben Nevis area. Although this area is hilly at present and even mountainous locally, at the time of emplacement of the granites at depth—Devonian times—it was probably very mountainous. Likewise the Cornwall and Devon granites of Hercynian age occupy the core of an old mountain range which extended in an east–west direction across those areas nearly 300 million years ago. The range has now been dissected and nearly completely destroyed by subsequent erosion processes. Most of these Hercynian masses such as Dartmoor have a surface area well over 40 sq. miles, but every so often off-shoot masses occur with outcrop areas much less than this figure. They are called **stocks** if their outcrop is irregular or **bosses** if it is roughly circular.

Whatever the size of these deep-seated bodies it is usual to find around the margins a zone of baking of the country rocks much wider than in the case of shallower intrusions. This zone or **metamorphic aureole** may reach widths of three miles or more, depending on the shape and size of the intrusion and such factors as magma temperature and composition. The junction between

PLATE 1

Volcanic breccia associated with a vent cutting through Lower Carboniferous rocks, East Lothian

PLATE 2

Columnar jointing in Lower Carboniferous basalt lava flow, Fife

PLATE 3

Spheroidal weathering of Carboniferous basalt, Shropshire

PLATE 4

Breccio-conglomeratic layers interbedded with coarse pebbly sandstone, Permo-Triassic, North Devon

a batholith and the surrounding rocks is often sharp but in some shield areas it is much more obscure. In some instances there appears to be gradation from the intruded igneous rock into the surrounding metamorphosed country rocks. Where the junctions are sharp, as is the case with all the granite masses of south-west England and the Lake District, an interesting effect can be seen. This is the inclusion within the marginal parts of the granites of fragments of the country rocks. The fragments vary from a few inches to many feet in size and are called **xenoliths** (Fig. 7). They have been wedged-off or **stoped**-off by the mechanical action of the fluid magma during its movement through the rocks.

The coarse and even-grained texture common to batholithic rocks which indicates very slow cooling is disturbed by the presence of xenoliths. It is also frequently disturbed by large feldspar phenocrysts. The Shap granite at the south-eastern end of the Lake District is well known for its distinctive porphyritic texture created by the presence of 1 to 3 in.-long phenocrysts of pink orthoclase. Some of the white feldspar phenocrysts in parts of the Dartmoor granite are even larger—up to 7 in. long. As in the case of the Shap the polished rock has been used a great deal for decorative purposes.

Veins

If a large granite or gabbro outcrop is examined carefully, there can often be traced very coarse-grained veins running not only through the igneous rock itself but also into the adjacent country rocks. These veins are generally 6 in. to 15 ft. wide and are called **pegmatites.** They suggest a special phase of deposition from highly-mobile hot magmatic fluids. The fluids are believed to have been released from the depths during the last cooling stages of igneous intrusion. Within the veins grains of mica have been found with diameters of 12 ft. and in North Carolina one grain found weighed nearly 2 tons. Feldspars up to 5 ft. long and quartz grains 3 ft. long are common occurrences in some North American and Brazilian pegmatites. Grains as large as these are very valuable and are used mainly in the electrical and ceramic industries. The pegmatites in the British Isles unfortunately are not so spectacular, neither are they so valuable economically as those overseas. Apart from common minerals the veins sometimes contain concentrations of rare but important minerals. One of the many uranium minerals, pitchblende, occurs in the Congo and Canada. Elsewhere wolfram, a source of tungsten, and cassiterite (the tin oxide) are found.

Pegmatites are not the only kind of veins associated with

igneous activity. In Devon and Cornwall there is a large number of north-east to south-west trending veins or **lodes** containing metallic ore minerals. The lodes are much more extensive in areal distribution than the pegmatites, and extend in a ten-miles-wide belt south-westwards from Tavistock in the east to St. Ives and Land's End. They cut through the granites and surrounding shales and sandstones and in some cases have been worked to depths of nearly a mile without entering into granitic rock at all. The mineralizing fluids exploited pre-existing lines of weakness in the crust such as joints and faults. The mineralized fault breccias are much more expensive to work because of the high proportion of rock waste.

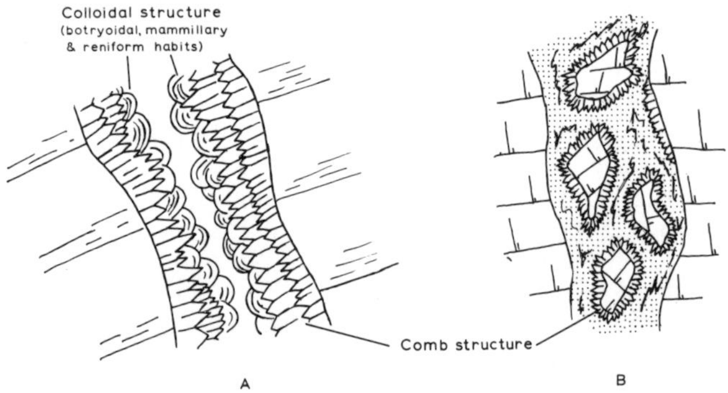

FIG. 8. A: fissure vein. Cavity infilling indicated by pyramidal termination of minerals. B: breccia vein associated with mineralization along fault planes. Contains a lower proportion of economic minerals

An important feature of the south-west England veins is that the nearer they approach the granites the greater is the concentration of high-temperature minerals in them. Within and immediately adjacent to the granites they contain tin and copper minerals. Those further away contain lead and zinc minerals and the most distant contain iron in the form of pyrites and siderite. The last were precipitated at a lower temperature than the others. **Mineral zoning** of this kind strongly suggests a connection between mineralization and igneous rock emplacement. On the other hand, in the Carboniferous Limestone areas of Derbyshire, the Mendips and northern Pennines the lead veins seem to have formed without large-scale igneous intrusions. No positive evidence of the presence of contemporary batholiths has been found. Consequently some geologists think that it is not essential to have an igneous magma source for large-scale mineralization or the development of mineral zones.

In the Carboniferous Limestone areas the galena and sphalerite were mainly precipitated in joints and fault planes. The lodes also occur as 'flats' following bedding planes in the limestones which have been partly enlarged by solution. The galena and the sphalerite are the ore minerals proper and they are intermingled with the gangue minerals fluorspar, barytes, witherite and calcite. Gangue minerals are not waste. Fluorspar, for instance, is used as a flux in iron and steel foundries and barytes is used in the manufacture of some paints. In lower-temperature veins the ore and gangue minerals commonly show **comb structure** (Fig. 8), in which the minerals develop good crystal shapes facing into the central parts of the vein.

Hydrothermal Activity

Towards the present-day surface of the land mineral veins usually show some effects of weathering by percolating ground-waters. For instance, the secondary limonite staining which characterizes the veins of Devon and Cornwall is due to the weathering of the primary pyrites and siderite. A change which was once thought to be due to surface weathering is the breakdown of granite into **china-clay,** a white soft rock mainly formed of the clay mineral kaolinite. Most geologists now believe that the alteration is caused by hot circulating fluids released during the last stages of cooling of a granite mass. The **hydrothermal** fluids have gases dissolved within them and attack and progressively corrode the feldspars and micas in the warm granite until only the original quartz and powdery kaolinite are left. A few grains of tourmaline and fluorspar are usually distributed through the clay and these plus the quartz have to be removed during industrial processing. The clay is a great economic asset in the export market and its main uses are as a filler in high-quality paper, a stiffener in textiles and as a raw material for the ceramic trade. Some of the largest and deepest china-clay pits are found around the south-western margin of the Dartmoor granite and within the St. Austell mass 30 miles to the west.

On other occasions a black variety of tourmaline called schorl is created on a large scale at the expense of the feldspars and micas without kaolinite forming. The products of this kind of hydrothermal alteration are equally as distinctive as china-clay. They include **luxullianite** which consists of partially-corroded pink feldspar phenocrysts set in a dark blue-black matrix of tourmaline and quartz. **Schorl rock** is a dark grey-blue admixture of tourmaline and quartz only, all the feldspars and micas having been eliminated.

Finally there are times when the hydrothermal fluids are abnormally rich in the element fluorine. Once again the feldspars and primary micas are progressively destroyed but instead of replacement by kaolinite or tourmaline there is the production of topaz (a fluorine-bearing mineral), fluorspar and large amounts of secondary mica. The new mineral assemblage plus the original quartz is called a **greisen.** Greisens are relatively commonplace in the south-west England granites but the best known, though somewhat topaz-deficient, occur at the margin of the Skiddaw granite at Grainsgill in the Lake District.

Although granite batholiths show the most spectacular effects of hydrothermal attack it should be realized that other types of igneous rocks can be similarly altered. Gabbros, basalts and ultra-basic rocks such as peridotites are all susceptible to drastic breakdown and very complex new rocks are produced.

Joints and Tors

One of the characteristic features of larger igneous masses, as with smaller ones, is jointing. Granites commonly exhibit at least two major near-vertical joint systems. This results in a crude columnar structure and some of the columns may have a diameter of several feet if the joints are widely spaced.

There are many other joint systems developed within cooling or cold masses but perhaps the most interesting are **cross-joints** which run roughly parallel to the present surface of the land. If the land surface is horizontal then the joints are horizontal; if it slopes to the north they slope to the north, and so on. They are probably caused by a release in pressure near to the surface as the rocks are eroded. Their main effect, however, is to divide the igneous body even further into near-horizontal slices. It is then criss-crossed by at least three sets of joints and it is the weathering along these which eventually gives rise to upstanding knobs of rock called **tors** in granite country. Tor formation is also probably assisted by patchy kaolinitization along joints and within the body of granites. The soft clay is stripped away very easily leaving behind the solid cores as tors.

Classification of Igneous Rocks

In the preceding sections the emphasis has been on the mode of occurrence of igneous rocks. All studies of rocks must begin by examining them in the field. In that way the field relationships are established, provided exposures are sufficient. Many problems, however, cannot be overcome by field studies and these include things like the chemical composition of the rocks and mineralogy of the fine-grained lavas. It is then that geologists enlist the aid of instruments such as the petrographic microscope. Using it the nature and relationships of the minerals in the rocks can be clearly established. Resulting from this combined outdoor and indoor work it is now realized that igneous rocks can be divided into separate groups in many different ways.

So far mention has been made of extrusive, minor and major intrusive modes of occurrence. Instead of using those terms some geologists prefer to use **volcanic, hypabyssal** and **plutonic.** They are possibly better as they suggest the conditions under which the rocks formed. Volcanic indicates surface activity such as lavas and volcanoes with rapid cooling effects. Hypabyssal, meaning shallow depths, refers to structures such as dykes and sills whereas plutonic suggests great depths, slow cooling and the production of coarse-grained lopolithic and batholithic rocks.

Because of the slow cooling plutonic rocks usually have grains larger than 1 mm. in size. These are easily seen in hand specimens. In contrast the volcanic rocks have cooled so rapidly that the grains are smaller than 0·05 mm. and are very difficult to see. Only if their texture is porphyritic can any individual minerals be really seen and these, in any case, are only the phenocrysts, not the ground-mass minerals.

Rock classification on the basis of field occurrence or grain size is not entirely satisfactory for the geologist's purpose. He needs some additional method which can be used to subdivide the rocks into even smaller groups. In order to help him out of this difficulty which first arose over a hundred years ago the geologist turned to the chemist and many analyses of rocks were made. The information gained from these chemical analyses proved ideal for the purpose and resulted in a subdivision based on the silica (SiO_2) percentage. Silica was selected because it was found to be an important component in all rocks. Those having a silica percentage greater than 66 were from then on referred to as the **acid** group, between 55 and 66 the **intermediate** group, between 45 and 55 the **basic** group and below 45 the **ultra-basic** group. The use of this grouping is illustrated in Fig. 9 into which the common igneous rock types have been fitted.

Even so many geologists are not completely satisfied with a classification based on chemical analyses. They say that the chemical composition of a rock cannot be determined by inspection in the field. Hence, in their opinion, the actual mineral content is the most important basis for classification. Their system has the added advantage that the minerals can often be seen. Fortunately the acid to ultra-basic terms can still be retained. The most distinctive feature of the acid group is its felsic mineral assemblage of quartz, feldspar and muscovite. Biotite is the only dark mineral which is at all common. The intermediate rocks are rather darker owing to an increase in mafic minerals, e.g. hornblende and augite. They increase at the expense of quartz and muscovite which become very minor in amount and importance.

	Acid	**Intermediate**	**Basic**	**Ultra-basic**
SiO_2 (%)	66	55	45	
Volcanic	Rhyolite	Trachyte Andesite	Basalt	
0·05 mm.				
Hypabyssal	Micro-granite		Dolerite	
1·00 mm.				
Plutonic	Granite	Syenite Diorite	Gabbro	Peridotite

FIG. 9. Igneous rock classification

The basic and ultra-basic rocks are darker and heavier still. This is not entirely due to the abundance of ferro-magnesian minerals such as augite and olivine but partly due to the incoming of the black iron oxides magnetite and ilmenite. The only felsic mineral of any importance in these dense, dark rocks is plagioclase feldspar and even this becomes insignificant in amount in the ultra-basic varieties.

SEDIMENTARY ROCKS

Sedimentary rocks are produced from the physical and chemical breakdown of pre-existing rocks. These rocks can be igneous, metamorphic or earlier deposited sediments. The materials derived from the breakdown in the source areas are transported in many ways. Some are carried by streams into lakes and seas in the form of boulders, pebbles and small mineral grains. Other materials enter into solution and are either used up by organisms in the formation of their skeletons or are precipitated directly from solution by chemical processes. In the case of organic life deposition of the skeletons occurs at death and they may accumulate in sufficient quantity to form a distinctive sediment.

Hence, sedimentary rocks can be subdivided into three main groups each of which has a fundamentally different mode of origin. Those that are composed of derived pebbles and smaller particles are called **clastic.** Those in which organisms play an important part are **biochemical** and formation by precipitation gives us the **chemical** group. Unfortunately some sediments do not fit easily into one or other of these groups. This is because mixed influences seem to have been at work in their production. Consequently, rocks may consist of a mixture of clastic, biochemical and chemical materials. **Bioclastic** limestones are a good example. Nevertheless within most sediments it is possible to detect a dominance of one kind of material and on the basis of this dominance the rock can be classified.

One of the most important diagnostic features of sediments in the field is **bedding.** The individual beds, bounded at the top and bottom by near-parallel planes, may be a few inches or several feet thick. The bedding planes probably represent short breaks in deposition; occasionally they are absent but only under special circumstances to be mentioned later.

Clastics

As these rocks consist mainly of derived boulders, pebbles and grains, the simplest form of subdivision into groups is on the size of the components. The boundaries between the groups are arbitrary but it is usual, at present, to make the 2 mm. average grain size the dividing point between the coarse pebble or **rudaceous** group and the finer-grained sandy or **arenaceous** group. The division between the sandy and finer **silty** groups is taken at $\frac{1}{16}$ mm. and between the silty and fine clay or **argillaceous** group at $\frac{1}{256}$ mm.

Rudaceous

Sediments of this type are found in many environments. People who are acquainted with mountainous areas are well aware of rock-falls and -slips which result in the blanketing of the lower slopes of hills with rock fragments. Wastwater, in the English Lake District, has around its margins spectacular wedge-shaped deposits of accumulated fallen debris. The topographical feature created is a **scree** or **talus slope** but the rock is more correctly referred to as a **breccia** (Fig. 10). The most important property of recent and fossil breccias is the angularity of the boulders and pebbles. This is hardly surprising as the fragments have not been transported any great distance and have not had the opportunity to rub against each other so as to become rounded. The lack of excessive movement accounts for the wide range in size of the fragments. It is not unusual to find pebbles an inch or so in diameter intermixed with boulders several cubic feet in size and weighing many tons. Breccias, therefore, commonly show **poor sorting.**

As soon as fallen fragments come under the influence of active running waters and are transported for some distance the sorting improves and most of the softer components become rounded. Indeed, some of the soft rocks may be completely worn away. The rock formed is a **breccio-conglomerate** (Pl. 4). Good examples of

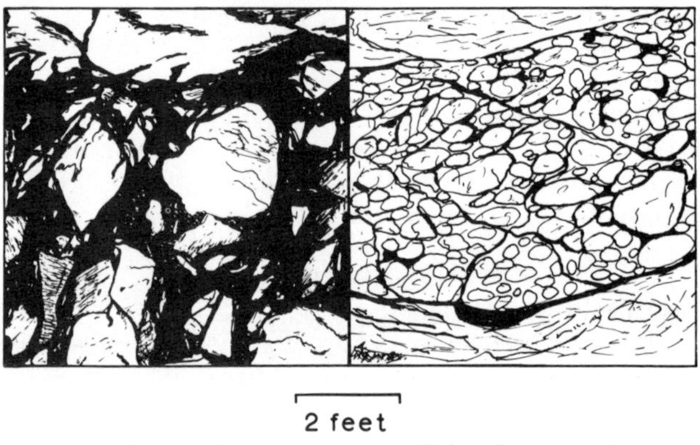

FIG. 10. Fragments in breccias are highly angular compared with conglomerates, and the sorting in breccias is very poor. Some conglomerates are very well sorted and even cross-bedded. Compare this figure with the volcanic breccia of Pl. 1 and the breccio-conglomerate of Pl. 4

this are the Brockrams of the Vale of Eden which consist of a mixture of angular and rounded slate, limestone and igneous pebbles. The Brockrams were formed during Permian times by the erosion of a mountainous land area to the west and south-west of the Vale.

The final product of fluviatile transport and deposition is a **conglomerate** (Fig. 10), in which nearly all the fragments are well rounded. Some of the Old Red Sandstone rocks of the Welsh borderlands and the Pentland Hills are of this type.

Not all conglomerates are formed by fluviatile transport and abrasion as can be deduced by inspecting many present-day beaches. Along parts of the south coast of England and especially near to chalk-with-flint cliffs there are beaches composed almost entirely of flint pebbles. The rounded flint pebbles have been derived from the adjacent chalk cliffs by normal marine erosion and the high degree of roundness is due to sustained abrasion caused by movements along the shore-line. Many of the more extensive fossil conglomerates are generally believed to be marine shore-line in origin rather than fluviatile. The basal Lower Cambrian beds of north Wales and the Tertiary Blackheath Beds of the London area are of this type.

Some boulder deposits on present-day continental shelves have a rather peculiar distribution as they occur in extensive patches well away from the land. The patches are up to 300 miles off-shore and are completely surrounded by finer-grained materials. A recent idea is that the boulders may have been transported from the shore by powerful and dense currents.

Density or **turbidity currents** (Fig. 11) are so called because they consist of pebbles, sand and mud all mixed up in a turbid mass. Being heavier than the surrounding clearer waters the mass hugs the bottom of the sea during movement. Speeds up to and occasionally well above 30 miles an hour are attained by the mass and may be sustained for many miles. Naturally, turbidity currents tend to move faster down the steeper submarine slopes and follow valleys and canyons carved in the continental shelves. In fact parts of submarine canyons seem to be intermittently filled with debris then swept clear by succeeding currents. Some geologists think that the canyons which reach depths of 1,000 to 3,000 ft. and are very steep-sided may have been created by the continual passage of turbidity currents, but their views are very controversial. A major difficulty with turbidity currents is visualizing how the mass movement of material ever commences. Possibly the loose unsorted sediment piles up towards the top of submarine slopes close to the land and is then triggered-off by slight earth-tremors.

The distribution pattern of boulders and pebbles on shelves may have nothing to do with turbidity currents. It is well known that melting ice-sheets and icebergs deposit morainic debris on the sea-bottom. The irregular distribution of pebble-, sand- and mud-banks in the North Sea (Fig. 15) is thought to be due to material dropped from extensive ice-sheets which moved outwards from Scandinavia during Pleistocene times. Large pebbles of Norwegian origin dredged up in fishermen's nets confirm the source. Turbidity currents could have had nothing to do with this as similar pebbles are also found in **boulder clays** deposited on the land in Yorkshire and East Anglia.

Boulder clays (or **tills**) are probably the best-known variety of terrestrial conglomerate. As the name suggests, the rock consists of boulders and pebbles set in a somewhat sticky, fine-grained clay matrix. Many of the fragments are derived from locally outcropping rocks but others may have travelled great

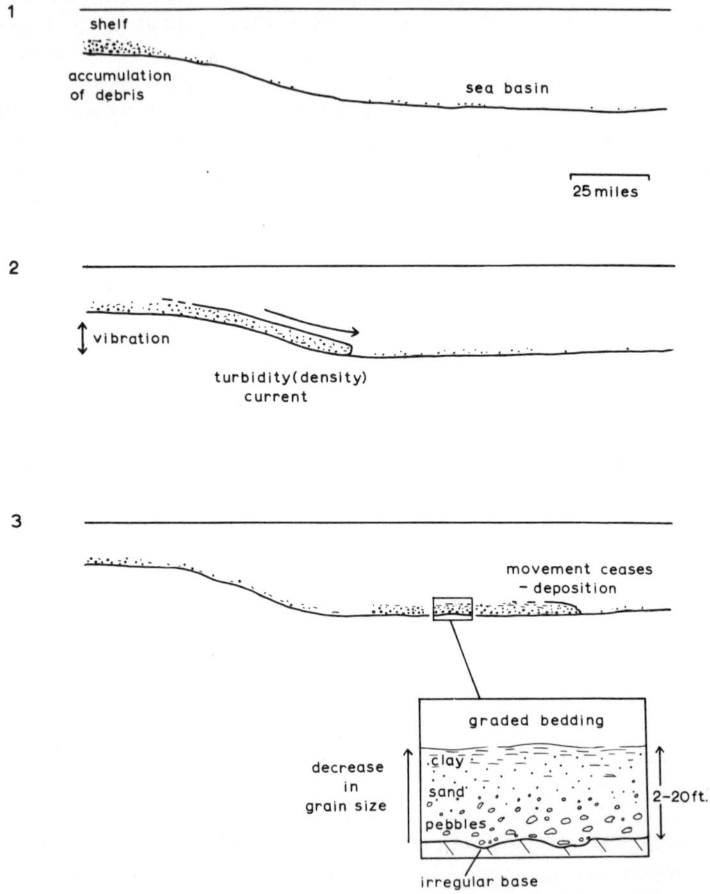

FIG. 11. Sequence of events leading to the production of density currents and graded bedding. Greywacke-type sandstones may be produced in this fashion

distances. Those which have been displaced far from their outcrop are called **erratics.** Large tracts of the British Isles northwards from London are mantled by boulder clay and in those areas it often prevents geologists from detecting the nature of the older rocks underneath. The clay is normally referred to as 'drift' on geological maps. In regions which have been glaciated more than once as a result of the advance, retreat and re-advance of ice-sheets two or more boulder clays may be found. The assemblage of rock fragments in each may differ considerably, indicating changes in the direction of ice-advance.

The final type of rudaceous rock which claims attention is a **fault breccia.** It is found occupying a zone of dislocation in the earth's crust. The width of a fault zone is quite variable, from a few inches to several feet, depending on the amount of movement and the nature of the country rocks. If the rocks are hard and brittle a good breccia forms; if soft and plastic no breccia is produced.

Arenaceous (including silts)

Sandstones, like conglomerates, are deposited in a wide range of non-marine and marine environments. They have, therefore, a broad range in composition, texture and structure. Despite this variation they can be grouped systematically on the basis of their mineral content and texture; structures are secondary in importance. Fig. 12 illustrates the grouping:

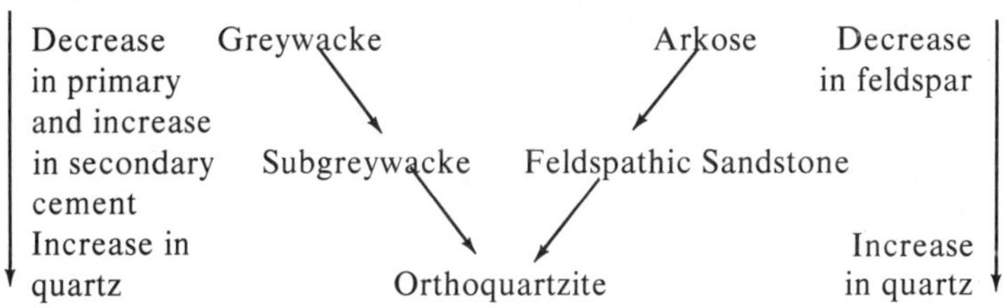

FIG. 12. Sandstone classification

The **greywacke** variety of sandstone is dark-coloured, black to grey in hand specimens. Sometimes a greenish tinge is seen and this is caused by a green flaky mineral called chlorite. The most characteristic feature of the dense rock is its texture of sub-angular to sub-rounded grains of quartz, feldspar and rocks set in a matrix of fine-grained minerals (Fig. 13). The matrix assemblage is not only of chlorite, quartz and feldspar but of micaceous particles called clay minerals (p. 37). Many geologists believe that the matrix is primary, that is, laid down at the same time as the large clastic grains. They suggest that the poor sorting exhibited by the rock, the high proportion of soft micaceous minerals and the angularity of the larger grains all indicate a lack of winnowing and little abrasion. Therefore the materials must have been deposited very rapidly, allowing no time for sorting

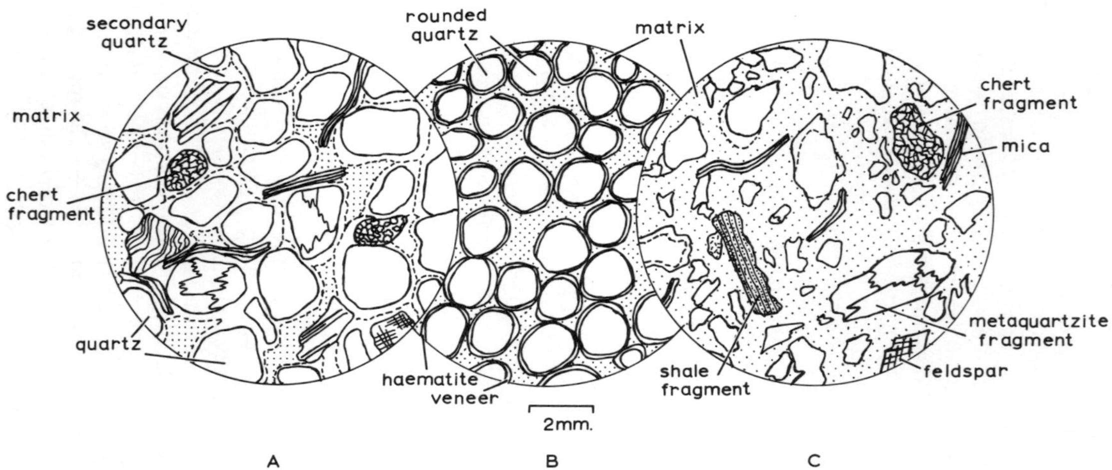

FIG. 13. Sandstone textures. A: Orthoquartzite texture. B: Millet-seed texture characteristic of wind-blown sands. C: Greywacke texture with angular grains and high proportion of primary cement

or active abrasion, and this may well have been below wave-base. A possible mechanism which would account for the texture could be deposition from turbidity currents and many people subscribe to this view. Caution is needed, however, as experimental work has shown that turbidity currents do not always create greywacke-type textures. Indeed, in some instances, a well-sorted sediment is produced. As a turbidity current origin is open to some doubt other ideas have been put forward. One is that chemical precipitation of clay materials from the water occurred simultaneously with the deposition of the large clastic grains. Another suggests that most of the matrix minerals are secondarily produced from the breakdown of feldspars and rock fragments a long time after deposition.

Greywackes are usually found in areas where there has been abnormal crustal downwarping. These are called **geosynclines** (p. 65) and within them many miles of marine sediments have been laid down. Sandstones may only form a quarter to a half of the total thickness, the remainder consisting of shales, siliceous deposits and igneous rocks of many kinds. In the Silurian greywackes of the Southern Uplands which were laid down in the Lower Palaeozoic geosyncline **graded bedding** structure is developed (Fig. 11). Each individual sandstone bed has a coarse-grained, sometimes pebbly, base which passes upwards progressively into finer-grained material. At the top of the bed there may even be a thin fossiliferous shaly layer. The mode of origin of graded bedding can also be explained in terms of turbidity currents carrying material in bulk both large and small. As the dense current slows down and finally comes to a halt the coarsest particles are deposited first because they are the heaviest. Then the finer and finer debris settles as the waters become less and less agitated. A vital point which must be emphasized, however, is that not all greywackes show graded bedding and, to repeat an earlier note of warning, graded bedding is not always restricted to greywacke sandstones. Bioclastic limestones sometimes show it.

Arkoses are highly feldspathic sandstones produced by the rapid erosion of a granitic mountainous area. They are deposited within and marginal to mountain belts. A large part of the source mountains were eroded away to form the Pre-Cambrian Torridonian arkoses of north-west Scotland; the main mountainous area to the north of Scotland has by now disappeared under the Atlantic Ocean. Similarly, only remnants of the mountains which gave rise to the Millstone Grit sandstones of northern England still remain in northern Scotland. Very few of the Millstone Grit beds are genuine arkoses, however, since they only contain 5 to

15 per cent feldspar. A true arkose should have more than 25 per cent. Most of the feldspars in arkoses are fresh and angular which confirms rapid erosion of the source rocks and indicates rapid transport, deposition and covering up by later sediments. This is because most feldspars are physically and chemically unstable and rapidly disintegrate if any of the above processes take place slowly.

The **orthoquartzite** group of sandstones, in contrast to greywackes and arkoses, is characterized by extensive re-working and sorting mainly on the shallow parts of continental shelves. The resultant sandstones are clean-washed and almost free of clay minerals and soft rock fragments. Only tough igneous, metamorphic and sedimentary rock grains are likely to be present, intermingled with the dominant quartzes and a few feldspars. The well-sorted grains are usually quickly cemented either by secondary quartz or calcite (Fig. 13).

Quartz cementation produces a very hard but brittle rock useful for road metal and refractory purposes. Such are the Wrekin (Cambrian) and Stiperstones (Ordovician) Quartzites of the Welsh borderland. The cement originates in various ways. Some is derived from the solution of the shells of dead siliceous organisms, e.g. sponges, diatoms and radiolaria, trapped between the sand grains. The silica from the shells is re-precipitated in the form of secondary quartz around the grains and welds them together. Other quartz is created by the partial solution of clastic quartz grains followed by immediate re-deposition of the quartz around the grains. Then it is possible that some primary inorganic quartz is precipitated directly from water. Primary inorganic calcite cement can also be precipitated from water but most is probably formed by the solution of the shells of calcareous organisms caught up in the sediment.

There are many other cementing agents such as haematite, siderite and limonite. Reddened sandstones so common to the Old and New Red Sandstone areas of Britain are rich in haematite precipitated under shallow water and terrestrial oxidizing conditions. As with many cements subsequent percolating ground-waters have caused a partial alteration and much of the haematite has now altered into limonite. Limonitization is a very common secondary cementation process in sandstones and gives the rocks a characteristic brown tinge.

The type and amount of cementation is very important when deciding whether a sandstone is a good source of water or oil. If the sandstone is poorly cemented and well sorted the **porosity** may be high, likewise the water and oil content. If, moreover, the pores are freely interconnected then the **permeability** will also be high and the fluids can be obtained from the pores by pumping if necessary. Some highly porous rocks may not be very permeable unfortunately and thus have limited economic value. Joints have an important effect on permeability of course, and good open jointing may compensate for a low permeability through the body of a rock.

Although the more areally extensive orthoquartzites are produced under shelf environmental conditions there are others formed under more specific and localized conditions. **Ganister** is one such variety found in coal-bearing sequences of fluviatile and deltaic strata such as the Upper Carboniferous and Middle Jurassic of northern England. It occurs as a pure, clean upper layer of a less pure sandstone bed and sometimes has a coal seam resting on top of it. Rootlet traces are always present. From this information it can be deduced that ganisters are part of old soil profiles created on sandbanks by colonies of actively growing vegetation. The vegetation appears to have leached out many of the impurities within the top layers of the sand.

In wind-blown terrestrial environments, especially hot and cold deserts, the winnowing action of the wind breaks down or removes the lighter and more friable micaceous minerals out of sand and creates a very pure rock. The Penrith Sandstone (Permian) in the Vale of Eden is an aeolian orthoquartzite. It also shows **millet seed** texture (Fig. 13) due to a high degree of rounding of the grains combined with evenness in grain size.

Whether deposited on land or in water orthoquartzitic sandstones are characterized by a large number of sedimentary structures. **Ripple marks** (Pl. 5) are very common on present-day sandy beaches and underwater photographs show that they continue to form out to sea at depths of 100 fathoms or more. In fossil sandstones they are equally abundant. **Cross-, current** or **false bedding** (Pl. 6) is the name used when describing sandstone beds which are cut by a series of closely-spaced planes oblique to the true bedding planes. These planes dip at angles of 5 to 30 degrees to the true bedding and, if observed closely, can be seen to be not completely planar. At the top and bottom they curve over until they parallel the true bedding. Each cross-bedding plane is thus divided into three parts. At the top and bottom there are the **top-** and **bottom-sets** and in the middle is the **fore-set.** By measuring the direction of maximum dip of the fore-set planes it is possible to work out the direction from which the clastic material came. This is because the fore-sets dip away from the source area. If a sand is being laid down by active currents or wind action it is usual to find that many of the top-set beds are missing owing to erosion.

Erosion can have a more startling effect. This is particularly so in a delta with its numerous drainage channels (**distributaries**). As the distributaries meander over the delta plain they cut down into earlier-deposited sediments and create new channels. Within them river-borne sandy and silty debris progressively accumulates until they become choked and the streams divert elsewhere. The choked channels are called **wash-outs** or **cut-and-fills** (Pl. 7). In the coalfields of Britain wash-outs often bring about difficult mining problems. If a rich seam is being worked and it abruptly terminates against a wash-out a decision has to be made whether or not to continue through, possibly, several hundred feet of barren sandstone.

Towards the base of some Pleistocene to Recent wash-outs and within some modern beach sands concentrations of rare and valuable minerals have been discovered. These localized deposits are **placers.** Much of the gold discovered over a hundred years ago in Alaska, Canada and the former Gold Coast occurred in placers. The largest nugget panned out of the Klondike sands and gravels in north-west Canada weighed 85 troy ounces. At Ballarat in Australia a nugget was found which weighed 2,268 troy ounces. In the Witwatersrand area of South Africa gold is mined from what some geologists think is a Pre-Cambrian placer, although others think the gold has a hydrothermal origin. Cassiterite (tin oxide) is another economically valuable placer mineral and it is dredged up from fluviatile sands and gravels in Malaya and the East Indies.

In addition to occasional rich pockets of heavy valuable minerals, all sands and sandstones contain 1 to 3 per cent of **heavy minerals** with no economic value. They are small in size and difficult to separate out from the rest but are very numerous in varieties. They include zircon, garnet, magnetite and tourmaline, all of which have a specific gravity greater than 2·80. The minerals are very important to the geologist because they are the most variable component within a sandstone. Two orthoquartzites, for example, with identical quartz content and texture may have completely different heavy mineral assemblages. Hence it could be deduced that the source rocks in each case were different even though the orthoquartzites were deposited under similar environmental conditions.

Inferences about source rocks based on the examination of heavy minerals come under the heading of **provenance study** (Fig. 14). The provenance of the Millstone Grit sandstones of northern England was a land area extending from north-west

Scotland to Scandinavia, formed to a large extent of acid igneous rocks rich in zircon and tourmaline and metamorphic rocks rich in garnet. Some of the Pliocene to Pleistocene Crag deposits of East Anglia contain heavy minerals which appear to have been derived from a Continental source. Many studies of this kind have been made over the last fifty years or so.

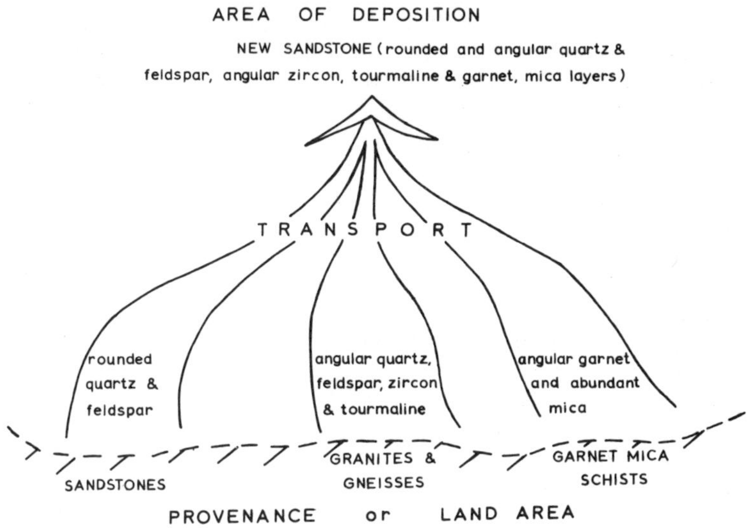

FIG. 14. Provenance study. The geologist infers the nature of the provenance by examination of the derived newly-created sandstone

Argillaceous

The sandstones grade imperceptibly into finer-grained siltstones (Pl. 8) then into **shales** and **mudstones.** The last two are true argillaceous rocks formed from the progressive consolidation of mud and clay. The consolidation is mainly attained by **compaction** of the original mud which drives out the large amount of water held in the pores between the mineral grains. Modern muds in estuaries and lakes contain more than 90 per cent pore-water, whereas Lower Palaeozoic shales contain about 3 per cent.

Grains of quartz, feldspar, mica and carbonates are widespread in argillaceous rocks but the most important, and usually dominant, minerals are the **clay minerals.** Modern work has shown that they are very small flaky minerals, not visible to the eye individually, and having a very complicated chemical composition. There are three important groups in which the varieties **kaolinite, montmorillonite** and **illite** appear.

The kaolinite group can be precipitated from and remain stable in slightly acid waters as might be found at the mouths of rivers. On the other hand, if the provenance is formed in part of kaolinite-rich rocks such as china-clays then the mineral is transported clastically and deposited irrespective of the chemical condition of the waters. Many non-marine and marine muds laid down adjacent to altered granite land-masses have a high proportion of kaolinite. The **ballclay** or china-clay deposits of Bovey Tracey in Devon are of this type. The kaolinite was eroded and transported from the hydrothermally-altered granites to the west during mid-Tertiary times.

Similarly the montmorillonite group is deposited clastically in a mixed series of environments if the provenance is rich in the mineral. It can also form by the alteration of mafic minerals under marine alkaline conditions. The mafic minerals (olivines, pyroxenes, etc.) may be transported from nearby igneous and metamorphic rock land areas or deposited directly in the water as volcanic ash. One of the products of this kind of alteration is a clay called a **bentonite** or **fuller's earth.** Fuller's earths occur in Lower Palaeozoic areas in Wales but the best known are Cretaceous and are worked near Redhill in Surrey.

The illite group is most abundant in shales older than Upper Palaeozoic but it is common in nearly all present-day marine and non-marine muds. The various members of the group appear to be stable whatever the chemical nature of the transporting and depositional water.

Because clay minerals are small and flaky and have a low specific gravity there is a natural tendency for them to be carried out much further from the shore-line than pebbles or sand grains. Many of the oceanic deeps of the world have a veneer of fine mud produced in this way. It is a mistake, however, to think that all muds are deposited well off-shore because of their constituent minerals' light weight. Within and adjacent to the mouths of rivers there is abundant mud deposition. Likewise, an equal error is made if it is thought that beach conglomerates grade systematically outwards from the shore into sands and muds. Conglomerates are often found many hundreds of miles off-shore. Marine continental shelves, in fact, are characterized by a most irregular pattern of sedimentation. A distribution-map of the North Sea sediments illustrates and confirms this important point well (Fig. 15). Similar patterns are found in large lakes such as the Great Lakes of North America.

A distinctive property of shales is fine **lamination** which divides the rock into very thin layers paralleling the bedding. Along these laminae the shale splits very readily into flat flakes—a property known as **fissility.** Mudstones, in contrast, are non-laminated and non-fissile. There is a coarser and less fissile lamination found in shales deposited in lakes in glaciated areas. These modified properties are caused by an increased silt and sand content. The coarse grains and mud are carried into lakes marginal

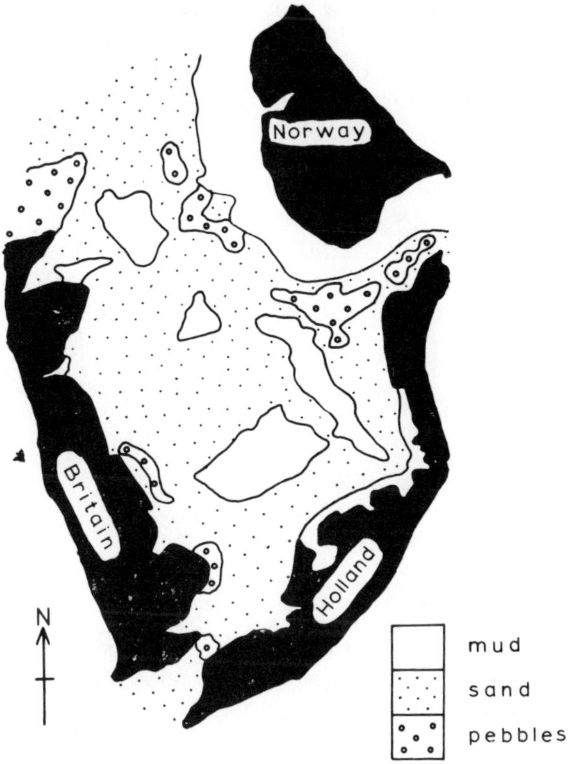

FIG. 15. North Sea sediments. The pattern is typical of shelf areas. Some of the pebble and sand deposits were probably originally laid down from ice-sheets in Pleistocene times

to glaciers and ice-sheets by warm season meltwaters. The silt and sand is deposited almost immediately on the bottom whereas much of the clay remains in suspension and only finally comes to rest in the cold season. Hence, each year sees a graded bed or **varve** (Fig. 16) created, with coarse material at the base passing upwards into shale at the top. Varves vary from $\frac{1}{4}$ in. to 3 ft. in thickness. As each cycle of deposition represents one year it follows that if present in quantity they can be used as a geological clock for dating contemporary events.

Most shales and mudstones are shallow-water marine. The grey-blue Oxford Clay (Jurassic) outcropping in a wide flat belt extending from Dorset to Yorkshire contains abundant marine fossils. It is actively quarried for making bricks along most of its outcrop. The sticky Gault Clay (Cretaceous) of south-east England is very similar. There are some argillaceous rocks where a completely different origin can be deduced. **Fireclay** is a highly siliceous clay with rootlet traces and is often found beneath coal seams. Although much softer and more plastic than ganister (p. 35) its origin must be very similar. The Carboniferous **oil-shales** of the Edinburgh area are highly bituminous shales formed in a lacustrine environment. The bituminous matter consists

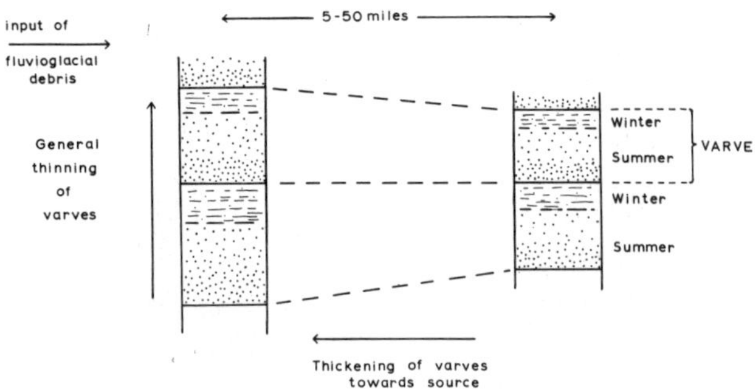

FIG. 16. The general trends in variation of varve thickness are caused by the slow retreat of the ice front

mainly of the waxy and resinous parts of plants and algal debris and it is these materials which give waxy oils when the shale is heated in a furnace. It is important to realize that oil-shales are not oil-impregnated rocks. The Scottish shales yield about 25 gallons of oil per ton of rock.

In humid tropical and subtropical areas rocks are intensively weathered at outcrop to form a type of soil and subsoil called **bauxite** (p. 78). These are high alumina clays which are valuable as a source of aluminium. If the clay is enriched in iron, as might be the case if basic igneous rocks are the parent rocks, then the name **laterite** is used. At present extensive bauxite and laterite deposits are being worked in the Guianas and Jamaica. Under the prevalent cool temperate climatic conditions in Britain no clays of this kind are forming but in the past the tops of basalt lavas of Carboniferous and Tertiary age weathered into laterites. In Antrim the laterites have been worked periodically. The importance of the British occurrences rests not so much in their limited economic value but in the way they indicate hot climates at the time of their formation.

Finally, there are a whole host of deep-sea marine muds or oozes about which little is known at present. Their clay mineral content is often quite low owing to dilution by organic debris raining down from nearer to the surface. Some of the muds have a reddish tinge but recent work suggests that most are grey.

Biochemical

The biochemical sediments are those in which organic life plays some part during their formation. Fortunately in most cases the organisms (animals or plants) which have helped to create the rocks are still present and visible. In a few instances they have subsequently disappeared as a result of lithification processes and a biochemical origin has to be inferred. Some bacteria appear to have the ability to make the chemical conditions suitable for the precipitation of iron and silica. As the rocks lithify the bacteria are destroyed or, at most, very poorly preserved.

Five biochemical groups can be distinguished, based on the nature and chemical composition of the organic remains. They are:

1 Calcareous
2 Siliceous
3 Ferruginous
4 Phosphatic
5 Carbonaceous

Calcareous

In tropical seas and warm oceanic currents microscopic calcareous organisms are abundant. There are many varieties of **foraminifera** with calcite shells and oozes formed of their shells cover nearly 50 million square miles of the ocean floors. Some of these forams are free-floaters near to the surface, that is, they are **planktonic,** but the majority are bottom-living or **benthonic.** In time the oozes consolidate into foraminiferal limestones (Fig. 20). The Chalk (Cretaceous) of Britain used to be thought of as a consolidated foram ooze deposited at great depths but it is now known that both these inferences are incorrect. In the first place forams are not abundant except at certain restricted levels. Secondly, the electron microscope, which is capable of magnifying grains 10 to 20 thousand times, shows that most of the Chalk is formed of minute algae (primitive plants) called **coccoliths** and **rhabdoliths.** Finally, it is now realized that limestone formation can occur in shallow, near-shore waters, providing little contaminating sands and muds are being brought in from the land. The Chalk, therefore, is a soft algal limestone laid down on a shallow-water shelf at depths not exceeding 100 fathoms adjacent to a low-lying land-area not being actively eroded.

The algae forming the bulk of the Chalk are disc-like, spherical and spindle-shaped whereas varieties found in **coral-reefs** are commonly net-like and encrusting in habit. This is very important in the growth of a reef because the net structure effectively binds together loose reef debris such as dead corals. In this way the reef structure is kept rigid and made capable of withstanding the constant attack of the sea. If the algae were not there the constant pounding by the waves would soon erode the reef and possibly kill off the actively-growing corals. Many coral-reefs have a higher algal than coral content, an odd condition which is duplicated often in older rock sequences. A typical biological analysis of a coral-reef is shown in Table 1.

TABLE 1

	per cent
Corals	34·6
Calcareous algae	42·5
Forams	4·1
Molluscs	15·2
Others	3·6

Data from Murray Island, Australia (after Pettijohn, 1949)

Biological analysis of a coral-reef

Reef-building corals and algae only thrive in mud-free marine waters at temperatures greater than 68°F. and with normal salinities. Oxygen and light are essential for the algae and some of the plankton which must be available as food for the coral polyps. Consequently the actively living part of any reef is the top 100 to 150 ft. The intermittently subsiding basal part of the reef, which may be many hundreds of feet thick, is dead but kept rigid by the elaborate intergrowth of its constituent organisms and extensive calcite cementation.

A coral-reef is an excellent example of a limestone formed 'in place'. A more technical name for this mode of formation is **autochthonous.** There are many kinds of reef and the current terminology is so complicated that only the simplest will be considered.

Fringing reefs (Fig. 17) are found immediately adjacent to the shore-line. When traced along the coast they may become separated from it by a lagoon which is sometimes hundreds of feet deep. The main reef body is then called a **barrier reef** (Fig. 17). Probably the best-known modern one is the Great Barrier Reef off the Queensland coast of north-east Australia. It is about 1,200 miles long, hundreds of yards wide and varies in distance off-shore between 20 and 150 miles.

Fringing and barrier reefs are found not only bordering large land-areas as in Australia and Arabia but also small oceanic islands a few miles across. There are many hundreds of these less spectacular reefs in the Bahamas of the North Atlantic and in the Pacific Ocean. In the latter the central island is often a volcanic peak formed of basaltic material. When there is no central island the oval to circular distribution pattern of the reef is then called an **atoll.** On the sheltered inner lagoonal side of an atoll, well away from the outer surf and active reef-growth

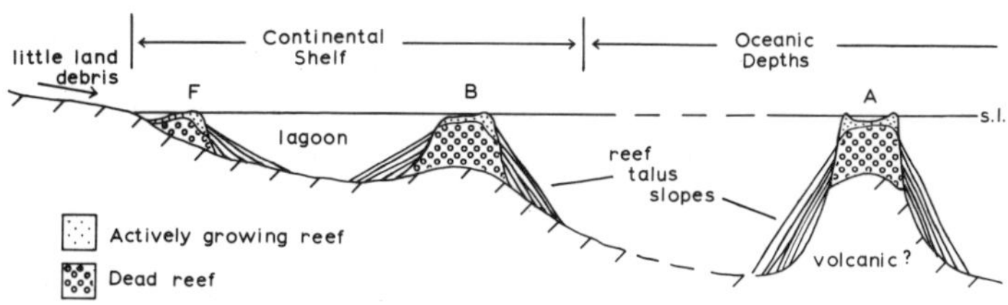

FIG. 17. Common varieties of coral-reef. F: fringing reef. B: barrier reef. A: atoll. In the Pacific area atolls often have a foundation of basaltic rock. Dead parts of reefs are usually dolomitized

zone, an accumulation of sand may progressively build up a low-lying land surface capable of sustaining vegetation and human activities. The sand is not a quartz-rich sediment but consists of sand-grain sized particles of limestone derived from the physical breakdown of the reef. It is thus strictly bioclastic in origin. On the seaward side of all reefs very extensive bioclastic submarine screes are formed owing to wave erosion. They are composed of angular coral-limestone boulders and pebbles and dip outwards at angles up to 40 degrees. Around some Pacific Ocean atolls the screes extend to depths of more than 300 fathoms. An important diagnostic feature of the screes is the presence of bedding planes. This contrasts with the reef proper which is generally devoid of bedding. When looking for reefs in older rocks, therefore, the geologist seeks firstly for algal-coral structures in which good bedding is absent and secondly for associated bioclastic breccias with bedding.

The earliest reef structures known are formed of calcareous

algae and are found in some late Pre-Cambrian sediments of Montana in North America. In Britain the oldest occur in the Wenlock Limestone (Silurian) of Shropshire; they are predominantly coralline. The most actively-studied reefs in Britain over the last eighty years or so are in the Lower Carboniferous areas of Derbyshire and the central Pennines. The great interest arises because they have all the usual structural features of reefs but are generally devoid of corals and algae. The only exceptions are the algal reefs of north-east Derbyshire. The apparent absence of genuine reef-building organisms is quite extraordinary. Some authorities account for this by saying that the original algae which created most of the reefs have had their filmy skeletons destroyed by later cementation and solution processes. The larger and more durable fossils were not affected.

As the innumerable varieties of fossil and modern reef habitats have been examined in detail it has become clear to geologists that no one hypothesis satisfactorily explains reef formation. Whatever hypothesis is valid for present-day reefs is not necessarily valid for Lower Carboniferous and other old reefs. Because of this every reef complex has to be judged on its own merits and not automatically linked up with ideas put forward by Darwin, Daly, Kuenen and many other famous scientists. The highly controversial hypotheses of these workers will not be discussed here.

Siliceous

The circumpolar seas, cold oceanic current areas and cold inland lakes are populated by an abundance of microscopic siliceous organisms allied to plants and called **diatoms.** Consequently, the bottom of these waters is characterized by the slow accumulation of **diatomaceous oozes** which consolidate to form **diatomite.** In warmer waters **radiolaria** and **sponges** may be almost equally abundant and in turn form oozes and eventually hard siliceous rocks such as **radiolarite.**

The Lower Carboniferous limestones of northern England and Lower Cretaceous sandstones of southern England often contain lenticles and nodules of siliceous rock in which sponge spicules and other organic debris is present. This is **chert.** Cherts are very widespread in old rock successions. They are not confined to limestones or sandstones by any manner of means but can be found in association with shales, ironstones and basic igneous rocks. One of the best-known nodular varieties of chert is **flint,** which occurs in the Cretaceous Chalk and in Tertiary to Recent conglomerates derived from its subsequent erosion.

Flint nodules are light grey to black internally with a white and somewhat powdery veneer caused by reaction between the silica of the flint and the calcite in the Chalk. Sponge spicules and silicified once-calcareous fossils are frequently present within them. The fact that organisms can be secondarily silicified is very important in deciding on the origin of the flint (and cherts in general). It suggests that many non-detrital flints and cherts are secondary in origin and were probably created a long time after the deposition of the enclosing rock. This is sometimes substantiated by nodules being discovered in joint planes and fault planes which are themselves clearly secondary. The source of the silica in secondary nodules is probably siliceous organisms distributed throughout the surrounding rocks. They are taken up into solution by percolating ground-waters and the silica is then chemically re-precipitated in the form of nodules or layers.

On the other hand the Ordovician strata of the Southern Uplands give strong evidence favouring a primary origin for chert. The cherts there are interbedded with shales and lava-flows. They are rich in radiolaria, well bedded, thick, areally extensive and do not occur in secondary joint and fault planes. Because they are also multi-coloured, but dominantly red, they are given the additional name of **jasper.** It is probable that the cherts record the accumulation of millions of dead radiolaria on the sea-bottom during a phase of non-deposition of muddy debris. Partial solution of the organisms subsequently helped to weld them into a compact, hard and brittle rock. In this lithification process silica released into the waters from the submarine lavas no doubt played some part.

Direct primary chemical precipitation of silica from fluviatile and marine waters, though doubted by many geologists, has been invoked for Pre-Cambrian cherts because of the apparent lack of siliceous organisms at that time. In the Lake Superior area there are thick cherts associated with ironstones which outcrop over many hundreds of square miles. Most of the chert is believed to be a primary precipitate of silica derived from adjacent low-lying lands formed of volcanic and other highly siliceous rocks.

Phosphatic

Extensive fossil phosphatic deposits such as the marine Permian **phosphorites** of North America are unknown in Britain. The Ludlow Bone Bed (Silurian) of the Welsh borderland and the Rhaetic Bone Bed in Gloucestershire are similar though much thinner and less areally extensive. They are similar because they consist of an accumulation of the phosphatic bones of vertebrate organisms such as fish. Many of the bony fragments have been rolled on the sea-bottom and have become nodular and well rounded and in this they have been assisted by secondary precipitation of phosphatic matter around the bony nucleus. The secondary process by which new phosphate is precipitated is called **phosphatization,** and apart from bony debris many calcareous fragments may be affected by it. Some of the calcareous fossils in the Cretaceous Gault Clay and Upper Greensand seem to have been prone to alteration during slight pauses in deposition on the sea-bottom.

Phosphate rock formation and phosphatization on land surfaces at present is confined to subtropical areas with spasmodic but heavy rainfall. Primary **guano** deposits are effectively leached under these conditions. Guano is found on islands and reefs colonized mainly by birds and it is the phosphatic part of their excreta which is partly taken up into solution and carried downwards into the underlying rocks. Strongly phosphatized andesites have been worked for fertilizer in the West Indies.

Phosphate deposition is probably very widespread on sea- and ocean-floors. Recent work has shown that about 6,000 sq. miles of the sea-floor off the southern Californian coast are covered by a thin spread of primary precipitated phosphate nodules and slabs. The nodules are concentrated on the tops of submarine banks and ridges between depths of 100 to 1,000 ft.

Carbonaceous

Carbonaceous matter is usually organic in origin although some is created by the high temperature and pressure alteration of carbonate minerals such as calcite. The main contributors in the production of carbonaceous matter are plants, especially when they have grown and died in abundance under swampy conditions. It is hardly surprising, therefore, to find that plant-rich successions are also rich in carbonaceous rocks. The Coal Measures of Britain illustrate this association very well and it is repeated in the Middle Jurassic beds of Yorkshire and Scotland and in some Tertiary beds of southern England and the Inner Hebrides.

In the Tertiary beds only a low degree of **carbonization** has affected plant concentrations and the resultant friable rocks are low-ranking **lignites** or **brown coals.** The older Jurassic and Carboniferous equivalents are hard black coals of higher **rank** (Fig. 18). They are usually finely laminated and micro-jointed but their most distinctive quality is a higher carbon content (about 85 per cent) and lower oxygen content (about 15 per cent) than that of lignite (carbon 75 per cent, oxygen 20 per cent). The ash content is also much lower.

Most of the pre-Tertiary household and coking coals are **bituminous** in rank. The top-ranking coals, however, are **anthracites** (carbon 94 per cent, oxygen 3 per cent, hydrogen 2 per cent) which are dull-grey to black, unlaminated, clean coals with very little ash content. The fact that coals vary in rank is difficult to explain. Age is an important factor. The older a coal is the more carbonized and compact it is likely to be but this does not explain why there is a considerable variation in rank within one specific coalfield. For example, the bituminous seams of the south Wales coalfield become anthracitic when traced towards the town of Ammanford on the northern side of the field. Depth of burial differences, chemical differences in the original plant

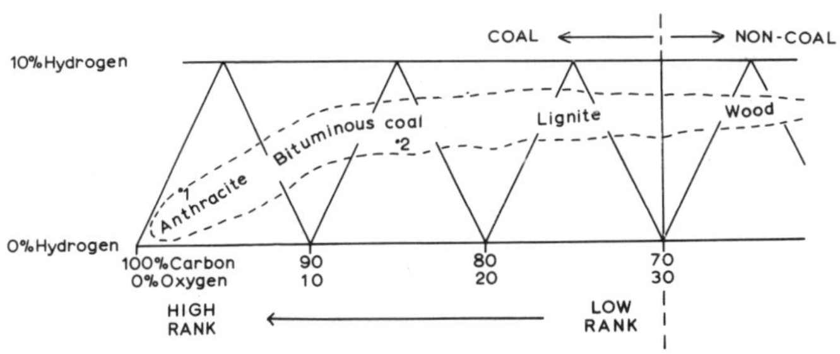

FIG. 18. Rank in humic coals. Coal 1 is anthracite (96% C, 2% H, 2% O). Coal 2 is bituminous (82% C, 6% H, 12%O)

debris and carbonization by large hidden igneous intrusions have all been suggested to account for the rank change. Some geologists believe that it may be due to earth-movement and low grade metamorphism. The more intensive the folding and faulting, the higher the rank of the coal. This certainly appears to be so in the coalfields of the United States where the anthracite seams always occur in the most highly-folded and faulted strata.

A significant characteristic of the lignite–anthracite coals is that they usually rest directly on a seat-earth. Some coals, however, do not have a seat-earth. One such variety is **cannel coal** which is a rather dull-black homogeneous rock occurring in the form of thin lenticular layers. It consists of plant debris plus a high proportion of clay all of which have been deposited in impersistent and more or less stagnant shallow lakes in the middle of swamps.

Whatever the origin of coal seams it is important to realize that they only form a low percentage—1 to 5—of the succession of rocks in any coalfield. In Britain they are interbedded with thick fluvial and swamp sandstones and shales with occasional marine shales. The seams are normally less than 10 ft. thick, commonly split into thinner layers or die out rapidly when traced laterally and are destroyed by wash-outs and subsequent volcanic activity. The last is a great problem in the Scottish Carboniferous coalfields.

Chemical

Chemical deposits are those in which the dominant process of formation is direct chemical precipitation. This precipitation can occur in fresh or marine waters and thus in a wide range of environments. One of the most important factors is that the waters should be free from any muddy debris and it is also advantageous if the rainfall in the area is spasmodic and the air and water temperatures are high. Under these ideal conditions the rate of evaporation at the surface of the waters is high and this concentrates the chemicals in solution so that they are more readily precipitated.

There are three important groups of chemical sediments, namely:

1 Ferruginous
2 Calcareous
3 Salts or Evaporites

Ferruginous

Most sediments (limestones, shales and sandstones) contain iron minerals but usually in small quantity. Only if the iron content rises above 15 per cent does the sediment qualify to be called ferruginous or an **ironstone.** An ironstone with such a low percentage of iron, however, is very poor in quality and would not normally be considered as an economic proposition. In the past some of the low-grade ores which have been worked include the Rhiwbina ironstone (Lower Carboniferous) of the Cardiff area, the Abbotsbury ironstone (Jurassic) of Dorset and the Claxby ironstone (Cretaceous) of Lincolnshire. The Coal Measures ironstones were also quarried extensively a hundred or so years ago. At weathered outcrops most of these rocks have a brownish tinge caused by the breakdown of the primary grey siderite, red haematite or green glauconite grains into limonite. Weathered ores are generally preferred by the iron and steel industry because their iron content is relatively higher than in the unweathered; much of the calcite has been leached out.

Higher-grade ironstones of Jurassic age are quarried in the Cleveland area of north-east Yorkshire, at Frodingham in northern Lincolnshire and in the Corby area of Northamptonshire (Fig. 19). The last is the richest ore with up to 35 per cent iron content. The most interesting feature of the ores is the range of rock types which constitute them. There are mudstones with a high siderite and chamosite (an iron silicate) content, sandstones cemented by siderite and oolitic rocks in which the

FIG. 19. Ironstones and evaporites. All are sedimentary in origin apart from the haematite deposits of the Lake District (Barrow-in-Furness)

ooliths (p. 51) and their cement are sideritic and chamositic. Calcareous shells of marine organisms are common at some levels and dilute the quality of the ore. The iron content also decreases as the rocks are traced laterally away from the localities mentioned. It seems likely that the ironstones were deposited in shallow marine waters into which abnormal amounts of iron were being introduced. North Yorkshire, Lincolnshire and Northamptonshire were near to the main outlets of rivers which may have been draining a low-lying area to the west formed, in part, of iron-rich Triassic sandstones and marls.

Although the richest sedimentary ironstones in Britain are Jurassic in age there are ores in the Carboniferous Limestone of Cumberland with an iron content of 50 per cent or more. The haematite ore occurs as irregular large pockets and lenticles and sometimes occupies post-Carboniferous fault planes. Hence, the ores must have been generated in the limestones after Carboniferous times. Most authorities believe the replacement of the carbonates in the limestones by haematite occurred in Tertiary times by upward moving warm hydrothermal fluids. Others suggest haematite replacement by cold circulating ground-waters carrying iron downwards from overlying haematite-rich Permo-Triassic sediments.

Calcareous

Probably the best-known calcareous deposits are **dripstones,** especially the stalagmitic and stalactitic varieties. Their origin is secondary and is caused by solution of the parent limestones followed by re-precipitation of the calcium carbonate. The rather impure calcite comes out of solution at a very slow rate because of the low temperatures prevailing in underground caves and it progressively builds up banded mounds, curtains and sheets.

Also of minor importance in limestone areas are **calc-tufas** which form irregular porous mounds and veneers on rocks where intermittent freshwater springs disgorge at the surface. Good examples are found in the Purbeck limestones of Lulworth and the Isle of Portland where the calc-tufa wraps around Jurassic tree stumps. The tufa seems to have been precipitated in very shallow freshwater swamps in which much vegetation was growing.

Whether it is deposited on a small or large scale, the most important factor in calcium carbonate precipitation is that the waters should be saturated with calcium and carbon dioxide. In warm oceanic currents there is normally a very high degree of saturation. The slightest variation in air and sea temperature or local current agitation can bring about chemical changes such that the waters become greatly over-saturated. A large-scale precipitation of calcium carbonate then occurs.

In the central shallow lagoons of atolls and on the shallower parts of some continental shelves fine-grained **aragonite muds** are accumulating at present. On Bikini Atoll in the Pacific Ocean and in the Bahamas of the North Atlantic aragonite muds can be detected which are being progressively converted into **calcite mudstones.** The aragonite is changing because it is an unstable form of calcium carbonate, while calcite is very stable. Owing to this instability very few aragonite mudstones are found in fossil limestone successions; calcite mudstones are commonplace. Some of the best examples occur in the Carboniferous Limestone of the Avon Gorge at Bristol. There, as elsewhere, it is usual to find algal and other organic debris intermingled with and cemented by the calcite mud.

Often closely interbedded with calcite mudstones and thus probably formed under similar conditions are **oolitic limestones** or **oolites.** Although shelly in part these consist dominantly of small spherical- to oval-shaped calcite particles called **ooliths** (Fig. 20). An average sized oolith is about 1 mm. in diameter. Some of the Middle Jurassic limestones forming the Cotswold scarps have ooliths of that size, but occasionally some layers have them up to 10 mm. in diameter. The name applied to these

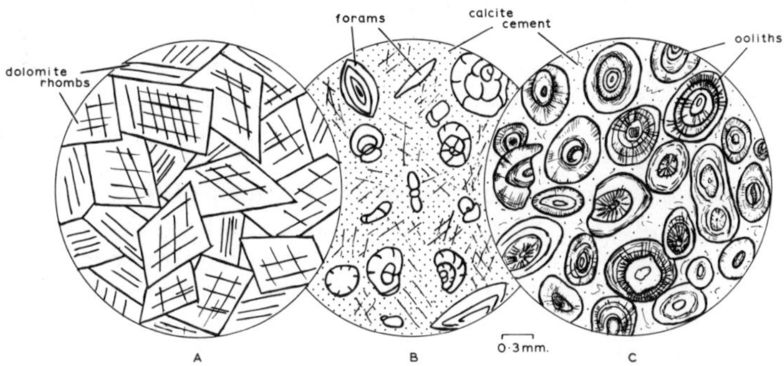

FIG. 20. Limestone textures. A: dolomitic limestone. B: shelly, foraminiferal. C: oolitic

rock layers is then modified to **pisolitic limestone** or **pisolite.** If ooliths are sliced open it can be seen that they have a concentric or onion-skin type of internal structure. This is caused by short pauses in the chemical precipitation of the calcite. At the centre there is commonly a shell fragment or quartz grain which has acted as a nucleus during the precipitation. Ooliths also show frequent evidence of abrasion which is a function of the fact that they are formed under shallow agitated water conditions. The limestones are sometimes strongly cross-bedded and exhibit wash-out structures.

A chemical process which afflicts all limestones whether they be coarse-grained oolites, fine-grained mudstones or are biochemical in nature is **dolomitization.** The process is secondary usually and either occurs on the sea-floor soon after the deposition of the limestone, or much later, after the limestone has been consolidated and uplifted by earth-movements. In both cases magnesium in the sea or in circulating ground-waters partially replaces the calcium in the mineral calcite. The newly-produced mineral, dolomite, is a calcium magnesium carbonate and has a characteristic rhomb shape when allowed to grow freely. The production of these rhombs can have a drastic effect on the original texture of a limestone, and in the end rock-product, also known as **dolomite** (Fig. 20), any ooliths or shell fragments may be completely destroyed. There are several excellent dolomites in the Carboniferous Limestone of south Wales and the Isle of Man.

If magnesium-rich fluids only partly alter an original calcite limestone the new rock is called a **magnesian limestone.** Original structures and textures are partly preserved in these in contrast to true dolomites. The Permian succession in Northumberland, Yorkshire and Nottinghamshire is noted for its magnesian limestones. Both they and dolomites are useful for refractory purposes in the iron and steel industry. Most, however, are converted into lime for building and agricultural purposes.

Salts and Evaporites

Apart from magnesian limestones the Permian system of north-east Yorkshire is renowned for its salt deposits (Fig. 19). The presence and wide extent (over several hundred square miles) of these beds was detected by a number of boreholes put down through a thick cover of overlying younger Jurassic rocks. Unfortunately, the depths at which the salt beds were found was so great that new techniques for working will have to be developed to make the recovery worthwhile economically. The salt beds are of various kinds. There are some consisting largely of rock salt (NaCl), others of anhydrite ($CaSO_4$) and gypsum ($CaSO_4.2H_2O$) and yet others of complex potassium and magnesium minerals, e.g. polyhalite. A feature of the beds is that they were laid down in a definite order or pattern which was repeated at least three times. At the base of each of the rhythmic units are beds of magnesian limestone. On top of them occur the gypsum-anhydrite beds and above these are the rock salt (halite), polyhalite and marly layers. It is clear that the chemical conditions bringing about the rhythmic precipitation of the minerals must have been repeated many times. The explanation is that northern England was occupied by a sea—the Zechstein Sea—during Permian times. This sea was almost entirely land-locked apart from an intermittent connection with more open seas to the east in Russia. Within it, and because of prevailing semi-arid, hot climatic conditions, chemical precipitation was the dominant process giving

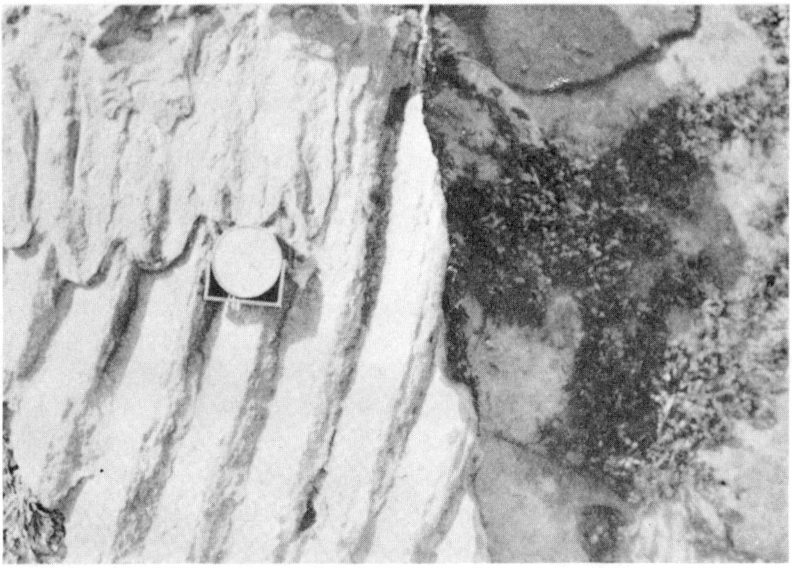

PLATE 5

Ripple marks in Lower Carboniferous rocks, Fife

PLATE 6

Cross-bedding in fluvioglacial deposits, West Ayton, Yorkshire

PLATE 7

Wash-out in Lower Carboniferous rocks, Crail, Fife

PLATE 8

Sun-cracks in siltstones, Fife

rise initially to the deposition of the least soluble carbonates, next the more soluble sulphates and finally the most soluble chlorides. At the termination of each of these cycles of desiccation, which probably saw a progressive decrease in the size of the sea, a renewed influx of marine waters occurred from the east. The whole process of desiccation was then repeated.

Climate is a very important factor in the formation of evaporites. High temperatures and a semi-arid climate, as in the region of the Dead Sea at present, mean a high rate of water evaporation. In inland lakes more water evaporates than is introduced by streams and rivers, thus concentrating the salts. The salts eventually deposited depend, naturally, on the composition of the waters. The Cheshire Plain during Triassic times was partly occupied by shallow lakes, the waters of which were dominantly rich in sodium chloride (Fig. 19). Consequently the evaporite layers are mainly composed of rock salt. Elsewhere in Britain the Triassic and Jurassic rocks contain nodules of gypsum and the Carboniferous Limestone a few miles to the south of Nottingham has gypsum-anhydrite beds.

Evaporite minerals are very important economically. The agricultural and chemical industries depend to a large extent on the processing of anhydrite for fertilizers and sulphuric acid. Potassium salts are also valuable as fertilizers. Hydrochloric acid is produced from rock salt, as is soda-ash, which is used in the manufacture of glass and soap.

METAMORPHIC ROCKS

Metamorphic changes are brought about by abnormal pressure, heat, chemical activity or a combination of all three. The new rock created differs either in texture, structure or mineral content from the original rock. In some metamorphic rocks all three changes have taken place whereas in others only one change, possibly textural, may be apparent. All kinds of rocks including earlier metamorphic rocks are capable of being changed, first by one process and then by another, until there is produced something which does not resemble the original in any way. One of the unfortunate effects of all this is the almost complete destruction of any fossil content present in the original rock. Metamorphic rocks are, therefore, very difficult to date accurately.

Although temperatures may rise to 600–700°C. during metamorphism it must be realized that the affected rocks do not ever become fluidized. At most they only become plastic. Another important point is that cementation and weathering changes are not metamorphic in the strict technical sense. Neither are dolomitization (p. 51), chertification (p. 44) and cold- or warm-water haematization of limestones (p. 50). Some geologists would say that anthracites and hydrothermally-altered granites represent special cases of metamorphism but they will not be regarded as such here.

Perhaps the most important generality about metamorphic processes is that they are all brought into being by earth movements. The more a region has been subjected to intense folding and faulting, the more metamorphosed the rocks are likely to be. It is no geological accident that the cores of past and present mountain ranges show a wide range of metamorphic effects varying from slight to very intense. Moreover, in these unstable areas large- and small-scale igneous activity is usually common and creates its own particular brand of metamorphism which may be superimposed on to the metamorphism caused by other processes. The complexity of effects is well demonstrated in the shield areas of the world, of which northern Scotland is one fragment.

If complex effects are unravelled it becomes possible to distinguish three kinds of metamorphism. These are:

1 Thermal or Contact (heat dominant)
2 Dynamic (pressure dominant)
3 Thermo-dynamic (heat, pressure and chemical activity)

Thermal (Contact) Metamorphism

This is best observed around the margins of igneous batholithic masses. The Dartmoor granite (Hercynian age) has a zone of baking or **metamorphic aureole** one to two miles wide. So has the less well-exposed Skiddaw granite (Caledonian age) in the Lake District. In each instance it is possible to trace a series of changes which become more pronounced as the granite margins are approached (Fig. 21). At the outer margin of the aureole the unaltered country rocks begin to toughen up and the shales in particular develop black spots. This is the **spotted slate** zone. The name slate can be somewhat misleading under these circumstances. The original shale is certainly hardened by baking but it does not always show **slaty cleavage** (p. 55) which is the most important property of a true slate. There are occasions, however, when the igneous intrusion post-dates the formation of slaty cleavage resulting from dynamic metamorphism. The spots progressively enlarge to form a greyish acicular mineral, chiastolite, nearer to the heat source giving the **chiastolite slate** zone. The new mineral produced is an alumino-silicate of square cross-section and the larger grains often have carbonaceous inclusions arranged in symmetrical patterns. At the margins of the granite the chiastolite zone in its turn is displaced by the **cordierite** or **hornfels** zone. Cordierite is an unstable silicate very difficult to identify in the field, hence the need to use the rock texture as a guide to position in the aureole. None the less cordierite (and chiastolite) are called **index minerals** because their presence does indicate a specific zone. Near to the contact almost complete re-crystallization occurs and a very tough but brittle rock, hornfels, is produced. Hornfels consists of an even-grained haphazard assemblage of quartz, feldspar, mica and cordierite.

As sandstones mainly consist of quartz, feldspar and mica in the first place, little change apart from toughening takes place

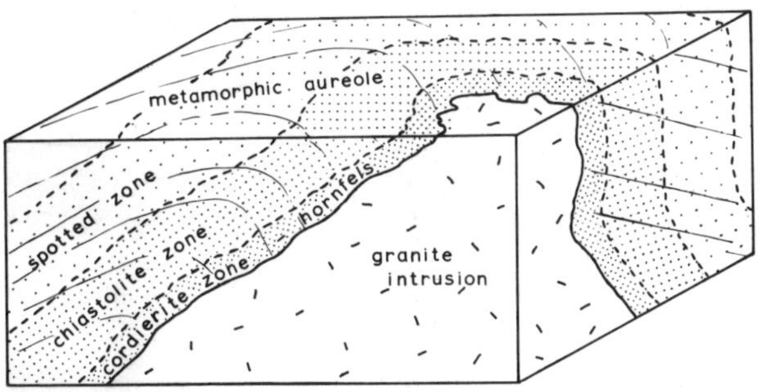

FIG. 21. Thermal metamorphism. The metamorphic aureole width depends on the form of the igneous intrusion underground

in them during thermal metamorphism. Limestones, in contrast, especially in the inner zones, recrystallize into **marble.** All fossil traces and primary textures are lost and a new closely-welded granular texture is produced. This hardened rock can be polished up for decorative purposes.

Dynamic Metamorphism

The best-known field expression of excessive dynamic processes is a true **slate** produced by the alteration of shales and mudstones. The main diagnostic feature of slate is a high fissility caused by the presence of closely-spaced parallel planes, about $\frac{1}{4}$ in. apart, known as **cleavage planes** (Fig. 22). These should not be confused with mineral cleavage planes. Often referred to as **slaty cleavage,** the planes commonly have a uniform amount and direction of dip over wide areas. In the Lower Palaeozoic slate belts of central and north Wales, the Lake District and the Southern Uplands the dip of the cleavage is near-vertical. One of the consequences of the formation of this secondary structure is the almost complete destruction of fossils. Preservation only occurs when the cleavage runs parallel to the original bedding. Usually nearly all the bedding planes in shales are destroyed, which makes it very difficult for the geologist to decide which way the beds are dipping. Fortunately, slaty cleavage is poorly developed in sandstones, limestones and lavas and if these rocks should be interbedded with the slates then the dip can be determined from them.

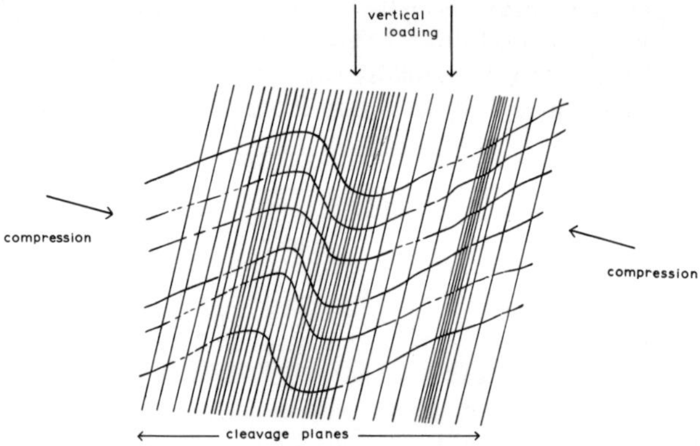

FIG. 22. Slaty cleavage. The traces of the original bedding planes in the shales can just be detected. The cleavage planes roughly parallel the axial planes of the folds

As soon as the direction of fold axes (p. 70) in a slaty area has been mapped, it becomes noticeable that there is a connection between folding and cleavage. The trend of the fold axes in fact roughly parallels the planes of cleavage. This suggests that folding and cleavage are generated simultaneously by compressive earth movements. The compression must have been at right angles to the axes and cleavage. Rocks vary in their response to compression. Shales are soft rocks and rich in flaky clay minerals which easily convert into other flaky minerals aligned at right angles to the lateral compression. Hence good-quality cleavage planes are produced. Most other rocks are harder and nearly devoid of primary flaky minerals; poor cleavage is usual.

It is significant that slates and interbedded rocks in mountainous areas do not show much evidence of change by high temperatures or excessive chemical activity. This suggests that really deep burial is not involved as appears to be the case with thermo-dynamic metamorphism. Purely dynamic effects are probably confined to the top 3 to 6 miles of the earth's crust during its acute deformation. The Lower Palaeozoic rocks of Wales and the Palaeozoic rocks of Cornwall and Devon show ample evidence of this high-level dynamic metamorphism.

Thermo-dynamic Metamorphism

The Scottish Highlands north of the Highland Boundary Fault are formed predominantly of Pre-Cambrian and Lower Palaeozoic rocks which have been metamorphosed on a large scale. The various new rocks created during several pre-Carboniferous phases of mountain building are exposed over a wide area. Because of this the name **regional metamorphism** is commonly substituted for thermo-dynamic.

If dynamic metamorphic rocks are created at high levels in the earth's crust during compression, then it is fairly certain that thermo-dynamic types are produced simultaneously at lower levels of more than 6 miles. The intense heat and pressure, many thousand times that existing at the surface, are generated partly by the original rocks being folded down into the depths and partly by hot chemical-rich magmatic fluids which permeate through the country rocks. Despite the abnormally high temperatures and pressures and chemical activity giving rise to rocks which lose all resemblance to the originals it must be re-emphasized that no fluidization occurs. The high pressures counteract any tendency of the original rocks to melt.

The greater the depth of burial and the closer the source of the hot fluids then the greater is the degree of metamorphism. Shales, for example, are converted into slates then pass into **schists** and finally **gneisses** as the metamorphism intensifies. Schists are very finely laminated or **foliated** rocks formed mainly of newly-created micas in which are commonly found new minerals such as garnet. Large new minerals are called **porphyroblasts.** The foliation is normally referred to as **schistosity** (Fig. 23). Gneisses are more coarsely foliated (or have a coarser schistosity) and generally consist of thin layers or flat lenses of quartz and feldspar alternating with mica. Intermingled throughout these minerals are newly-produced kyanite, sillimanite and other minerals. Kyanite is a pale-blue silicate and sillimanite is a fibrous silicate. They are strictly metamorphic minerals and can be used as index minerals for defining zones as in the case of thermal metamorphism.

Sandstones and acid igneous rocks are less changed than most rocks. The more quartzose sandstones are converted into tough

meta-quartzites and the less quartzose into slightly more foliated **granulites.** Granites generally become gneissose. Basic igneous rocks, especially those forming dykes and sills, convert into schists. Limestones being formed dominantly of unstable carbonates produce a whole host of new textures and rocks. With less intense alteration various kinds of marble are produced but more intense alteration creates tough, light-coloured, mottled rocks formed of a mixture of calcium and magnesium silicates. These **calc-silicate rocks** are completely devoid of all original carbonates.

Excluding Scotland, thermo-dynamic products outcrop at Start Point, Bolt Head and the Lizard in Devon and Cornwall. The hornblende mica schists and gneisses there are of unknown

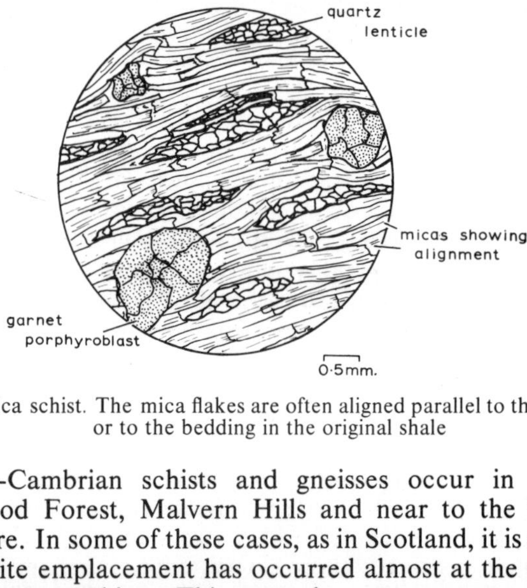

FIG. 23. Mica schist. The mica flakes are often aligned parallel to the lamination or to the bedding in the original shale

age. Pre-Cambrian schists and gneisses occur in Anglesey, Charnwood Forest, Malvern Hills and near to the Wrekin in Shropshire. In some of these cases, as in Scotland, it is significant that granite emplacement has occurred almost at the same time as the metamorphism. This strongly supports current ideas of the close relationship between mountain building, metamorphism and granite intrusion.

As thermo-dynamic changes and large-scale earth-movements are so closely linked, it follows that schists, gneisses and marbles can be formed at any period of time. Some of those outcropping in the Alps were formed in Tertiary times. In Brittany there are schists and gneisses created in mid-Carboniferous times and there is no doubt that some of the Scottish rocks were altered during the Devonian period. The tendency to consider all metamorphic rocks as being Pre-Cambrian, therefore, should be avoided.

QUESTIONS

1 Write a brief essay on coal and coal-bearing strata with particular reference to the Carboniferous rocks of Britain.
2 Distinguish between sedimentation on land-surfaces and that of shelf sea areas.
3 Why are sandstone studies important?
4 Describe the textures and structures within igneous rocks which indicate that they are of extrusive origin.
5 Account for the textures of deep-seated igneous masses and describe the hydrothermal changes which may affect them subsequent to emplacement.
6 What field evidence can be used to detect that an igneous body is discordant?
7 Briefly describe the connection between mountain building, granite batholiths and regional-scale metamorphism.
8 Define the term metamorphic aureole, then describe the changes that can be observed in shales when intruded by a granite mass.

Chapter 4

EARTH STRUCTURE

THE uppermost layers of the earth are widely exposed on land-surfaces and a large amount of information has progressively accumulated about them by direct observation. The same cannot be said about the much greater bulk of rocks forming the internal parts. The gold mines in the Transvaal are only 2 miles deep and the deepest boreholes for oil only 5 miles; they are mere scratches on the surface of a planet which has a radius of approximately 3,960 miles. Consequently the geologist has to rely on indirect evidence in building up a picture of the interior structure. In this he is assisted by many studies which have been made of earth-quakes.

Earthquakes

Earthquake tremors are due to movements in the earth's crust at depths up to many hundreds of miles. These may result from faulting, that is an actual physical displacement of rocks which may or may not be detectable at the surface. Alternatively, the tremors may reflect plastic flowing movement beneath the brittle crust. In many of the more disastrous earthquakes such as have afflicted parts of California, the Mediterranean lands, the Andes and the Himalayas in the past the physical effects have been very obvious. For example, during a 1906 tremor the ground 30 miles north-west of San Francisco was displaced horizontally for 20 ft. In south-eastern Alaska in 1899 an earthquake was accompanied by a rise of 47 ft. in the land-surface at one locality. Vertical and horizontal movements, in fact, are very common occurrences and often produce small **fault scarps** (Fig. 24). These should not be confused with **fault line scarps** which result from differential erosion of hard and soft beds on either side of an old fault plane.

Huge landslips, particularly in mountain areas, may be triggered-off by earthquakes. They are most dramatic, destroying everything in their path, especially if the material is channelled down valleys. Some of the most powerful and destructive mass movements of material occurred in the Quetta district of the north-west Himalayas in 1935 and in Chile in the early 1960's.

When submarine tremors occur not only are small fault scarps produced but also **tsunamis.** Tsunamis are seismic sea waves which travel at speeds up to 450 miles per hour away from the **epicentre** of the tremor. On board ship in deep oceanic waters very little sea swell is seen but as the waves cross the continental shelves they increase to a terrifying height. Around the coasts of Japan and eastern Asia waves of nearly 100 ft. in height have been recorded in the past 200 years and in the mid-eighteenth century an estimated height of 210 ft. was observed. The effects of these waves on low-lying coastal regions are equally disastrous as any

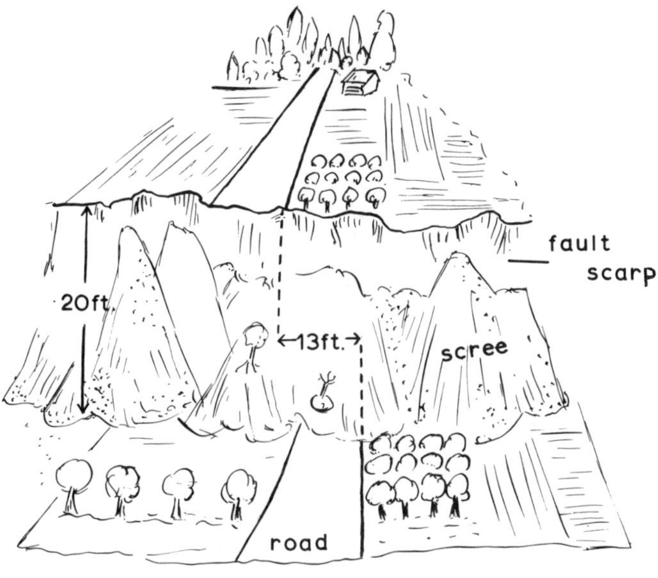

FIG. 24. Fault scarp formed during earth-tremors in Japan, 1891 (after Gilluly). Most major faults have a lateral and vertical displacement

land-focused earthquake. Many of the 60,000 people who perished in the 1755 Lisbon tremor died by drowning.

People who live near to active or quiescent volcanoes become used to regular tremors of small intensity. These disturbances are caused by the pressure of the fluid magma at depth. If the disturbances become more frequent and intense then safety precautions have to be taken because it is then fairly certain that the volcano is going to erupt. The island of Tristan da Cunha in the South Atlantic had to be evacuated quickly in 1961 when the tremors became very intense and hot lava began to be explosively ejected from vents.

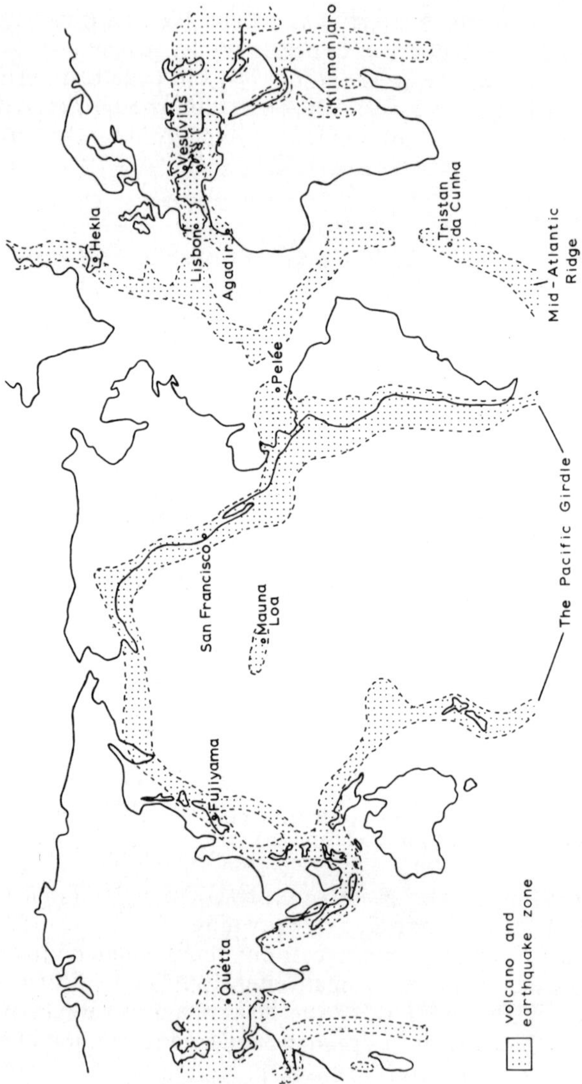

FIG. 25. Active earthquake and volcano zones of the world

The distribution pattern of active volcanoes (Fig. 25) and the foci of the most powerful earthquakes throughout the world indicates a very important relationship. This pattern coincides with the zones where Tertiary to Recent large-scale mountain building is found. The coincidence is not accidental. It shows that volcanoes and earthquakes are a phase or a surface indication of active mountain building. The Lower Palaeozoic areas of Britain must have been repeatedly shaken by vulcanicity and earth-tremors prior to and during their main upheaval into mountain ranges at the end of Silurian times. There is ample evidence of vulcanicity during Ordovician times but it is much more difficult to demonstrate contemporaneous earth-tremors. Perhaps the contortions of some beds resulting from their slumping down shallow sea-bottoms is such evidence. It is thought that submarine slump movements can be started by earth-tremors.

Deep-seated Structures

Earthquake intensities are measured by instruments called **seismographs.** The actual movements of the rocks are automatically recorded on a graph and the closer the tremor to the instrument the more intense is the movement of the earth and so the registration on the graph. By combining and comparing the readings from stations distributed over a wide area it is possible to fix the geographical position and depth below the surface of the focus of the earthquake. More important, however, is that the graph also indicates the difference in time it takes for different types of wave generated by the earthquake to travel from the focus to the recording station. Some of these waves (Primary P and Secondary S) pass through the deeper parts of the earth whereas others (Longitudinal L) pass along, near to and at the outer surface. The P and S waves are the most informative because they travel not only in straight lines outwards from the focus but also in zigzag bouncing fashion along the junctions between the various shells of the earth. The peculiarity of these reflected or bouncing waves which have taken a deeper and much lengthier route to the recording station is that they sometimes arrive much quicker than theoretically expected. This means that some of the deeper layers transmit waves much faster than the upper layers and it is a condition which could only arise if the rocks at depth are denser than those above.

From this earthquake information it is now believed that the earth consists of a series of concentric shells with increasing density towards the centre (Fig. 26). Below the veneer of sediments which rarely reaches a thickness greater than 8 miles, and then only on continents, there is a crustal layer which varies from 18 to 30 miles in thickness. Towards the top of this crust but beneath the sedimentary cover the P waves travel at approximately 3·5 miles per second. This is the speed theoretically expected if rocks similar to granite form the layer. As there cannot be any confirmation that granite is, in fact, present the term **sial** (si for silicon and al for aluminium) is used instead. The sial grades downwards without any sharp break into the lower crust where P waves travel at about 4·5 miles per second. The rock at this level is more like a basalt in composition and is, therefore, referred to as **sima** (si for silicon and ma for magnesium). An important point about sial is its absence from the greater part of some oceanic areas. Even the sima forming the floor of

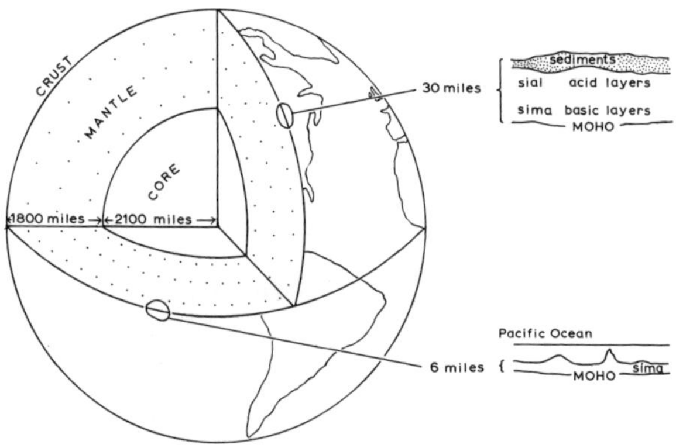

FIG. 26. Internal structure of the earth

the Pacific and Atlantic Oceans is only a few miles thick. Consequently there is much interest in deep-sea boreholes which might penetrate through the sima into the unknown underlying rocks.

The junction between the sima and the underlying **mantle** rocks is a very important one first detected by Mohorovicic, a Yugoslav. It is now named after him and is called the **Mohorovicic discontinuity** or 'Moho' for short (Fig. 26). The discontinuity occurs at a depth of about 30 miles under continents and a few miles under oceans and the rocks beneath it are characterized by a greater uniformity in physical properties than those above. The P waves travel at a more uniform rate of about 6 miles per second which suggests that the dense rocks are ultra-basic in composition.

The mantle is about 1,800 miles thick and as far as can be

inferred from indirect evidence has a relatively uniform solid composition throughout. The core (2,100 miles in radius), in contrast, is probably formed of a viscous fluid. This inference is based on the fact that some earthquake waves do not appear to be transmitted through it. As rigid solids transmit all waves it follows that the core must be in a different physical state from the rest of the earth, namely fluid. The nature of this viscous material is controversial but it must be of a very high density and most geologists think it is probably a dense mixture of nickel and iron.

Geosynclines

Geosynclines are long narrow sags in the crust created by large-scale deep-seated earth-movements (Fig. 27). Although superficially similar to synclines (p. 70) in form, their scale and mode of origin are so different that no confusion should arise.

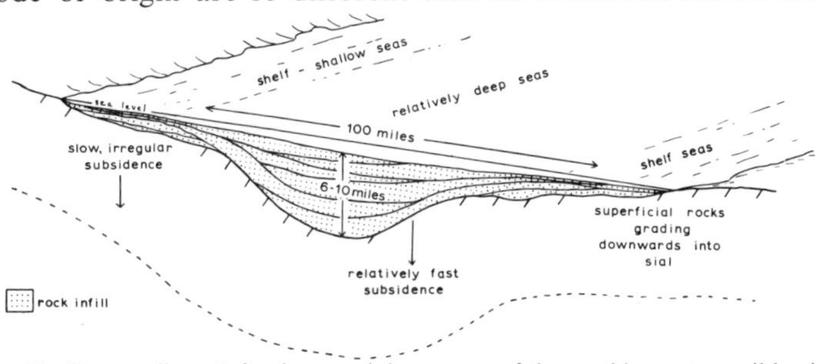

FIG. 27. Geosyncline. A fundamental downwarp of the earth's crust possibly aided by the loading effect of the thick pile of rocks laid down in the axial areas. Greywacke sandstones, shales, volcanic ashes and lava flows are prominent members of the rock infill. After compression during mountain building phases the width of the site of a geosyncline may be reduced by tens of miles

They are areas of abnormal rock thickness compared with adjacent more stable areas and are very susceptible to subsequent compression. A major geosyncline existed approximately on the site of the Mediterranean Sea in Palaeozoic and Mesozoic times. In its axial regions at least 10 miles thickness of sediments and igneous rocks accumulated and were uplifted in late Carboniferous and Tertiary times to form **fold mountain** ranges. The second phase of uplift resulted in the Alps.

In Britain a geosyncline occupied the site of the Irish Sea, Wales, the Lake District and southern Scotland during Lower Palaeozoic times (p. 154). Something approaching 40,000 ft. of sediment (at least twice the thickness found in the marginal shelf areas) was deposited in the north-east to south-west trending axial regions and later compressed to form the Caledonian Mountains during Devonian times.

Geosynclines can thus be regarded as the sites of future mountain ranges, the internal structures of which will eventually comprise many folds and fractures.

Outer-layer Structures

These are structures which can be inferred by observation of the rocks in the field. Clearly this observation only concerns the thin layer of rocks resting on top of the sialic and simatic shells. Nearly three-quarters of the rocks exposed on land surfaces are sedimentary; the remainder are igneous and metamorphic. The main structures resulting from igneous and metamorphic processes have been dealt with earlier so only those associated with sedimentary rocks need be examined.

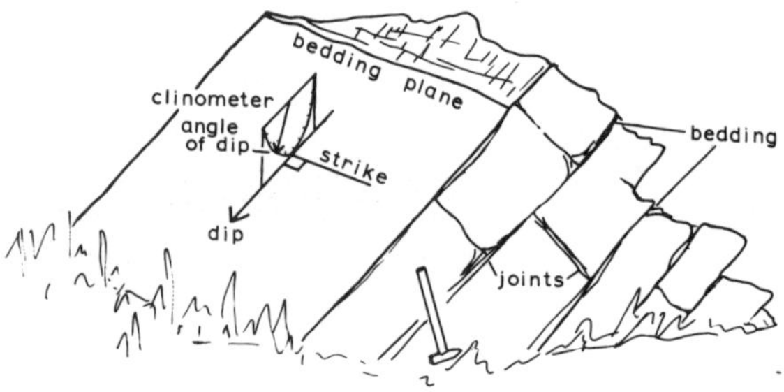

FIG. 28. Measurement of dip and strike

The most important structural feature of sediments is **bedding** or **stratification.** The top and bottom surfaces of each individual bed, i.e. the **bedding planes,** may be inches or many feet apart and are usually parallel to each other. The individual beds, sometimes uniform in lithology, sometimes not, rest one on top of the other and a reasonable deduction is that the lowest bed seen is the oldest and must have been deposited first. This generalization is called the **Law of Superposition** and is only untrue if the beds have been overturned by earth-movements. If a bedding plane is horizontal then it has no **dip.** The dip is the amount of inclination from the horizontal and is either recorded in degrees or in terms of gradient, e.g. 1 in 10. An instrument called a **clinometer** is used for either type of measurement. If a bedding plane is tilted to any angle up to 90 degrees then it has not only a measurable maximum dip direction but also a **strike** direction. The strike of a plane is always at right angles to the true dip (Fig. 28). The fact that the strike direction, by definition, has no dip means that on any bedding plane it is possible to construct a series of horizontal lines called **strike lines.** These are similar to contour lines on a topographical Ordnance Survey map, including the use of sea-level as datum point. The only difference is that strike lines are constructed on a bedding plane which usually disappears underground whereas topographical contours are drawn on the existing surface of the land.

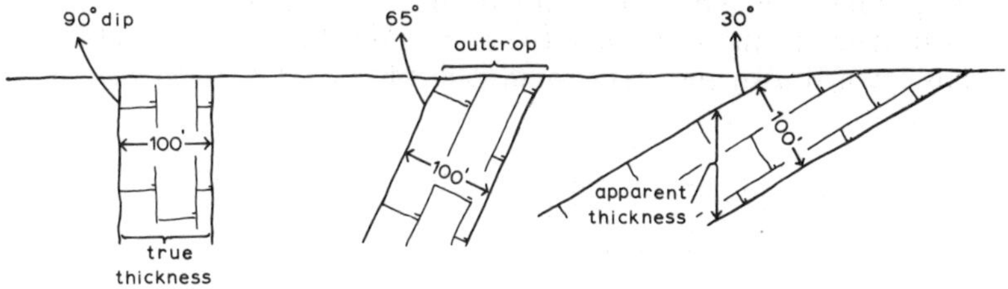

FIG. 29. Variation in width of outcrop for a given thickness of bed is dependent on the angle of dip. The lower the dip the greater the outcrop, if the topography is flat

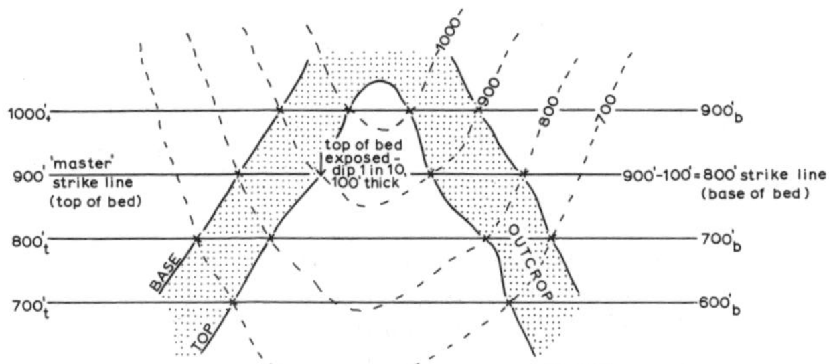

FIG. 30. Construction of strike-line systems and completion of outcrop. The 'master' is perpendicular to dip direction. Gradient 1 in 10. Therefore draw 100-ft. strike lines 1000 ft. apart parallel to the 'master'. Constancy of dip of the bedding planes is assumed

Beds which are exposed at the surface or, more commonly, just hidden beneath a thin cover of soil are said to **outcrop.** The actual width of outcrop depends on many factors such as the thickness of the bed and the topography. The angle of dip of the bed is also very important. For a bed of uniform thickness the width of outcrop decreases as the angle of dip increases (Fig. 29).

One of the problems of a field geologist is to map in the outcrop of the beds when the number of exposures is low. It is then that strike lines become of utmost importance because by using them in conjunction with other field data he can still map the outcrops with fair accuracy. What is the minimum amount of field information required for this operation? Clearly the thickness of the bed is important, as is the angle of dip. If the direction of dip is known as well then the geologist can construct his map. The manner of construction is illustrated in Fig. 30.

The direction of the strike lines is controlled by the direction of dip of the bed and the distance between them by the angle or gradient of dip. The computation of the strike-line spacing also depends on the scale of the map. After the strike-line system has been drawn on to the map it is completed by following the

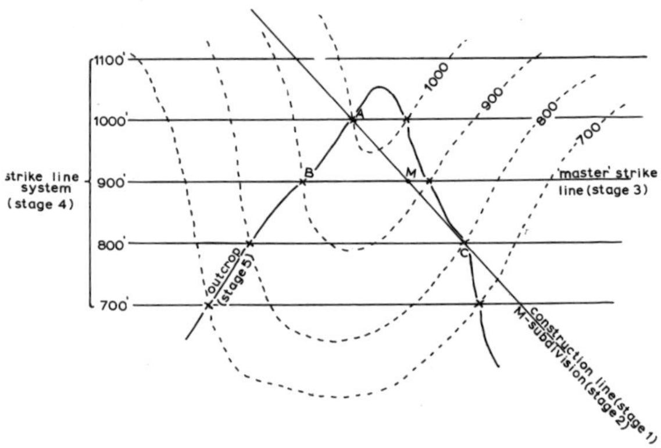

FIG. 31. Three-point construction. A and C are the highest and lowest fixed points on the construction line. M is the point on the construction line which coincides with the height of point B. Constancy of dip of the bedding plane is assumed

simple rule that the points of outcrop (bedding planes in this case) must appear at the surface whenever the height of the land coincides with the value of the strike lines. For example, if a 500 ft. strike line intersects a 500 ft. contour line, then at the point of intersection the bedding plane must outcrop. Having drawn in the outcrop pattern for one bedding plane which may be the top of the bed the outcrop of the base is then constructed by subtracting the thickness in feet of the bed from the original strike-line values.

Sometimes exposures are so poor that it is impossible to measure the dip of a bedding plane at any one point. Providing the geologist can detect the same plane at three different places, however, he can still work out the true dip (Fig. 31). This method is known as the **three-point construction.**

Unconformity

When beds in cliffs and quarries are examined closely it is usual to find that they all dip uniformly or concordantly in one direction. The rock succession is said to be **conformable.** There are occasions, however, when it can be observed that two groups of superimposed rocks are each dipping independently of the other. They have an uncomformable relationship and the junction between each group is an **unconformity** (Fig. 32). Sometimes it is rather flat, suggesting marine planation, as at the base of the Cambrian rocks in Shropshire; in other cases it is very uneven indicating terrestrial weathering and erosion, e.g. the base of the Permo-Triassic in south Devon (Pl. 9). Only the flat type of discordant junction can be called a **plane of unconformity.**

The significant point about any kind of unconformity is its indication of a break in deposition during which erosion and

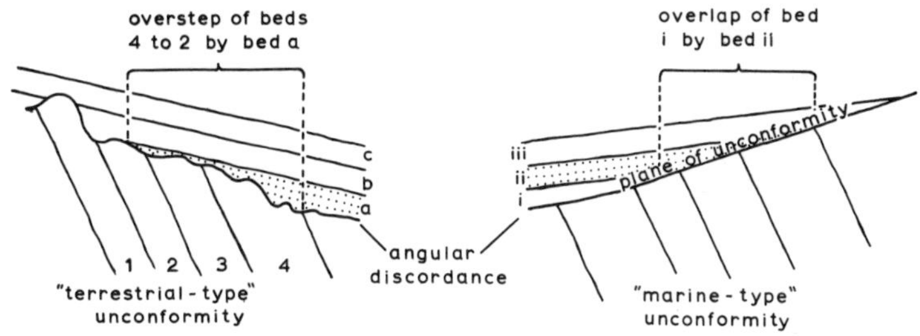

FIG. 32. Unconformities, overstep and overlap

folding have occurred. If the angular discordance between groups of beds is slight then the period of erosion, folding and, sometimes, non-deposition has probably been slight. In these instances the older rocks may not have been lifted above sea-level. If, in contrast, the discordance is a large one then it is certain that a major uplift above sea-level followed by intensive erosion has taken place. Unfortunately the actual unconformable contact between beds is usually hidden by soil creep, rock-falls and vegetation. Under these circumstances and despite the fact that a marked discordance in dip may be noted in the field the geologist has to look for further evidence of a break in deposition. If the beds are fossiliferous he might detect a marked change in the nature of the fossils between the two groups of strata. More useful, however, is the evidence afforded by the younger group because if it rests on different members of the older group it is almost conclusive proof of unconformity (Fig. 32). This diagnostic feature is called **overstep**; the younger group is said to overstep successive beds of the older. Almost of equal value is **overlap** (Fig. 32), a feature created mainly by marine waters impinging against and progressively transgressing across a land surface.

Folding and Faulting

Folding and faulting are nearly always the companions of uplift and erosion. Folding involves a bending of beds often in a complicated fashion whereas faults are generated when beds have insufficient elasticity to bend.

The simplest folds are found in those sediments which have not been subjected to intense compression or which rest directly on a rigid basement of old hard rocks such as granites and metamorphic rocks. The open folds formed under these conditions are **synclines** and **anticlines** (Fig. 33). Sometimes one **limb** may be steeper than the other giving a marked asymmetry. When

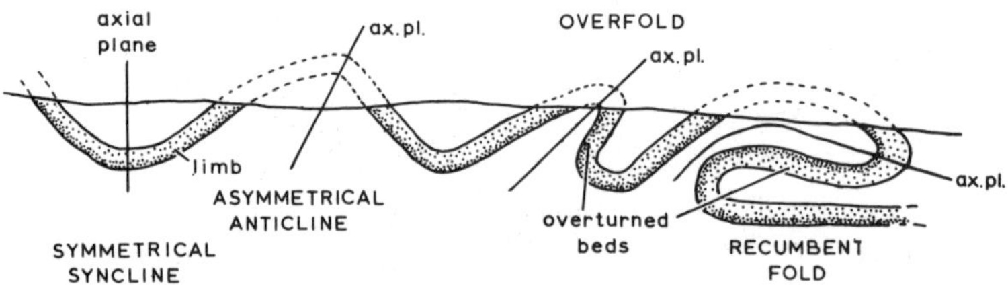

FIG. 33. Folds

the thickness of sediments is much greater and compression has been more intense a high degree of asymmetry is usual and the limbs of folds are often overturned. The inversion of one limb of **overfolds** and **recumbent overfolds** (Fig. 33 and Pl. 11) means that the Law of Superposition becomes invalid and great care is needed in working out the true succession of rocks.

If a particular bed is traced from one limb of a fold to the other there is a point where a minimum dip or even no dip can be detected. Through this point runs the **axis** of the fold. If the axis is horizontal then no dip is measurable along its full length. On the other hand, if the axis is not horizontal a dip can be measured which is known as the **plunge** or **pitch** angle. The effects of plunging or pitching axes on fold patterns is quite important because they make the outcrop of the beds converge or diverge in a V- or U-shaped fashion (Fig. 34). Good examples of this can be seen in the Mesozoic rocks at Henfield in Sussex and Poxwell in Dorset. At the latter locality the anticlinal axis plunges away at each end of the fold giving rise to what is known as a **pericline.**

Folding combined with subsequent erosion or erosion on its own often creates structures called **outliers** and **inliers.** An outlier is an outcrop of younger rocks completely surrounded by older (Fig. 35). Excellent examples of Lower Jurassic marlstone outliers occur down the western flanks of the Cotswold Hills in Gloucestershire. In contrast, an inlier consists of older rocks completely surrounded by younger. Along the eastern margins of the Derbyshire Dome at Ashover and Crich occur inliers of Carboniferous Limestone flanked by younger Millstone Grit rocks.

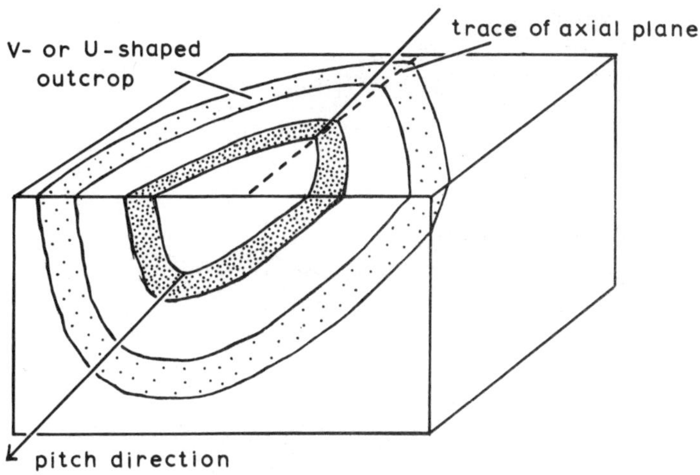

FIG. 34. A syncline plunging towards the viewer. Note the effects on the surface outcrops

Most areas of the earth's crust where slight or intense folding has occurred have invariably been affected by faulting. **Faults** are physical dislocations of the outer layers and are more common in areas of intense folding. The displacement on each side of the **fault plane** may be very small and measure only a few inches or feet but the physical effects none the less may be devastating, involving much loss of life. This is because the release of tension in the crust brought about by the fault movement may be accompanied by earth-tremors. In older faults displacements may measure many hundreds of feet, for example the Southern Uplands Fault which is the southern boundary of the Midland Valley of Scotland. Such extensive movements as this are unlikely to have happened at one instant of time. They are the cumulative result of innumerable small displacements similar to those observed at the present day in earthquake areas.

If the rocks on each side of a fault are hard and brittle a considerable amount of brecciation occurs as the rocks move past each other and a **fault breccia** is produced. The breccia may be many feet thick and is often secondarily mineralized. Sometimes the grinding action along the fault plane creates smooth but grooved surfaces; these are **slickensided** faces.

There are four main kinds of faults (Fig. 36). **Normal** faults (Pl. 10) are generated by tension or a pulling apart of the rocks.

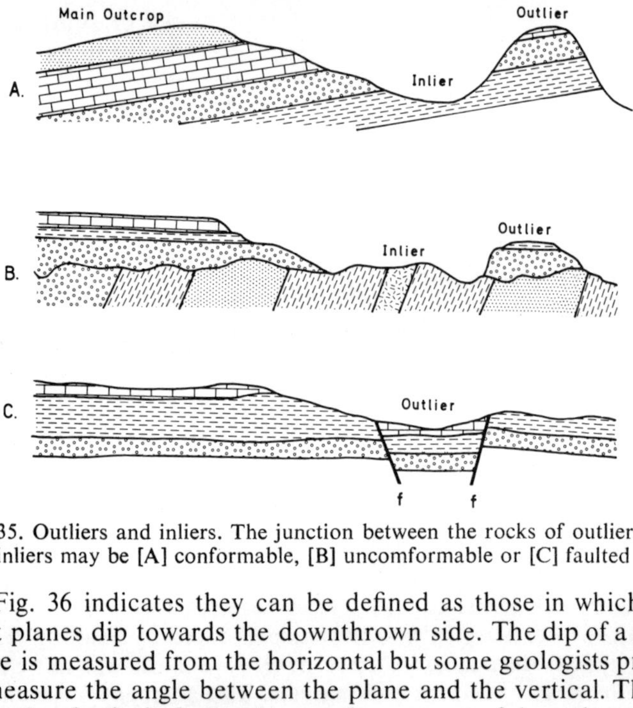

FIG. 35. Outliers and inliers. The junction between the rocks of outliers and inliers may be [A] conformable, [B] uncomformable or [C] faulted

As Fig. 36 indicates they can be defined as those in which the fault planes dip towards the downthrown side. The dip of a fault plane is measured from the horizontal but some geologists prefer to measure the angle between the plane and the vertical. This is the **hade** of a fault. In most cases the amount of downthrow can be calculated by comparing the displacement of a given bed on each side of the fault. Strike lines are useful in this process especially with problem maps. The amount of throw is the difference between the values of the strike lines for any bedding plane common to each side of the fault where they meet at the fault. Combinations of normal faults such as in the middle reaches of the River Rhine, in the Middle East and East Africa produce a structure called **block** and **rift** (or **horst** and **graben**) (Fig. 36).

Both parts are fault-bounded, the rifts having been downthrown relative to the bounding blocks on each side (Fig. 36). Rifts are commonly low-lying elongated depressions occupied by rivers and lakes and hence are also referred to as **rift valleys.**

Reverse and **thrust** faults (Fig. 36) result from compression. A reverse fault is one in which the fault plane dips towards the up-thrown side. The amount of dip is usually greater than 30 degrees. If the dip is less then the name thrust fault is used. In areas of active mountain building the earth pressures have often been so intense that the thrust fault planes are nearly horizontal. The Moine fault in north-west Scotland is a good example of this.

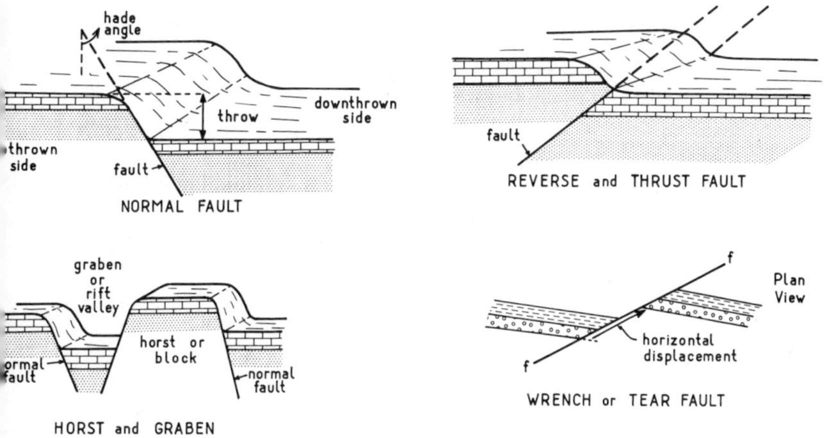

FIG. 36. Faults. Normal faults are caused by tension whereas the others are compressional in origin

The Pre-Cambrian rocks have been displaced bodily in a north-westerly direction over younger Cambrian rocks for a distance of at least 10 miles.

Wrench or **tear** faults (Fig. 36) are common in the South Wales coalfield and they result mainly from compression. They differ from the other types of fault because the displaced rocks have not moved upwards or downwards but sideways. No throw can be measured, only lateral movement. The Great Glen of Scotland is etched out along a very large wrench fault on each side of which the rocks have been relatively displaced about 60 miles.

Jointing

Fault planes should not be confused with **joint planes** even though they resemble each other in many ways. Both are usually vertical or oblique and both may be mineralized but the rocks on each side of joints show no displacement. Joints are much more common than faults and there are few exposures, natural or man-made, in which they cannot be detected. The origin of joints is controversial although most geologists recognize that they are partly due to the warping of consolidated beds. The release of pressure caused by vertical loading when the rocks are uncovered may also help to create them.

QUESTIONS

1 Describe and illustrate the following: (*a*) overlap, (*b*) overstep, (*c*) thrust fault, (*d*) horst and graben structure.
2 The earth consists of many shells. Outline the main features of these shells and explain how their presence has been detected.
3 How would you detect the presence of a fault in the field?

Chapter 5

LANDSCAPES

LANDSCAPE is the surface expression of the underlying rocks modified by surface effects such as climate and vegetation. Dynamic though almost imperceptible changes are taking place continually and may be observed at the present time. Freshly exposed rocks and land-masses are instantly attacked and, in the long run, obliterated, their material forming the basis for new rocks. The destruction of the surface of the land can be referred to the following processes: **weathering** or the breakdown and fragmentation of exposed rock, **transport** or the removal of the broken-down material and **erosion** or the wearing away of material. The last is a dynamic process often carried out by the agents of transport using tools provided by weathering. These processes work together and must be considered as complementary though it is convenient to separate them for the purpose of study. Closely linked with them is **deposition** which can create new features in the landscape from the material provided by the destructive elements.

The continuation of landscape formation depends on the instability of the earth's crust which constantly provides fresh surfaces and renews the available land-areas. Throughout geological time the earth's crust has been unstable. At the present day in areas subject to earthquakes and vulcanicity the land may rise or fall many feet over a period of a few days or even hours. In more stable regions, such as the Baltic Shield or Britain, movement is much less violent and tends to be a slow bending or warping taking place over tens of thousands of years. The Pre-Cambrian and Lower Palaeozoic rocks of Norway and Sweden have domed upwards many hundreds of feet during the last half a million years or so. This seems to be due to the melting of a huge thickness and weight of ice which previously had compressed downwards the underlying rocks. As the load on the rocks decreased so the land began to bend upwards to a position of greater equilibrium. The process whereby rocks move regionally and relatively slowly to restore and maintain equilibrium with their surroundings is called **isostasy.** One of the most obvious effects of isostatic adjustment is the occurrence of raised beaches (p. 99). These can be seen around many of the coasts of Scotland. It must not be thought that isostatic or other more drastic earth-movements always mean uplift of land above sea-level. Downwards movements are equally common. The coastal margins of East Anglia, for example, seem to be warping downwards towards the North Sea at a rate of about one inch every 100 years or so.

Weathering

Weathering takes place in a number of ways. The chief mechanical processes are insolation, frost action and the activities of animals and plants. The chief chemical processes are solution and recrystallization. The atmospheric conditions which bring about weathering also control its degree and type. In cold circumpolar regions or above the snow-line of mountains physical breakdown effects tend to be dominant whereas in subtropical and tropical low-lying areas physical effects are subordinate to chemical.

The effect of heat from the sun's rays is to make rocks expand. This expansion is an effect of **insolation** and varies from one group of rocks to the next because of the different rates at which the component minerals expand. The direction in which the rocks face is also important as the heating effect is more pronounced on the slopes directly facing the sun. The areas where the effect is most pronounced are the rockier parts of hot deserts. There the sun beats down fairly continuously during the day and then at sundown the heat gained is rapidly radiated away into the cloudless sky. The temperature range under these conditions may be well over 100°F. The cumulative effect of daily expansion and contraction within the outer layers of the rocks eventually brings about their decay and the layers flake off or **exfoliate** revealing fresh layers. They in their turn are then attacked. Not all geologists believe that insolation can produce extensive exfoliation. They think that chemical breakdown is more likely to produce the same effect even under desert-like conditions.

Frost action is more common in colder climates including mountain peaks where the temperatures are regularly at or below freezing-point. The process by which the rock fragmentation occurs is basically simple. Water percolates through natural fissures, joints and minute cracks during relatively warmer phases then freezes and expands as the temperature drops below freezing-point. The expansion of the ice wedges the fissures and cracks apart until a stage is reached where fragments are loosened and fall under gravitational pull to the lower slopes. Screes are a frequent result.

Animals, such as earth-worms, use rocks and the soils produced from them during their life-cycle. Their ceaseless boring and burrowing activities progressively churn and comminute the rock particles to the finest of clays. The roots of trees and other plants can be even more powerful agents of breakdown by physically wedging open rocks, especially along joints. The fractures are enlarged and the rocks fragmented further by the chemical attack of permeating organic acids. Live organisms secrete corrosive fluids but acids are also created from the breakdown of dead organic matter. Much chemical weathering is organic in origin, though circulating surface and ground-waters contain many inorganic chemicals which are more important in their effects. Carbon dioxide in solution acts as a very mild acid. When carbon dioxide-rich waters percolate through rocks formed wholly or in part of alkaline materials the rocks tend to be corroded and may disintegrate. Limestones and calcareous sandstones are particularly prone to this kind of attack. Most minerals are capable of being dissolved by natural waters. A few are more resistant to solution and either remain unchanged, e.g. quartz, or recrystallize. Anhydrite ($CaSO_4$) commonly hydrates or recrystallizes into gypsum ($CaSO_4.2H_2O$) and magnetite, haematite and glauconite hydrate into limonite.

Soils

The best-known product of weathering is soil. In areas where physical processes are dominant the soils tend to be thin and poor in quality. On the other hand, more than climatic factors enter into their production. On chalk and limestone, whatever the climate, cultivation is very restricted due to poor soil development. The parent rocks are composed predominantly of soluble carbonates which produce little or no residue on solution. Any residue left behind is composed of insoluble clay minerals and iron oxides which are thinly sprinkled throughout the original rocks. These progressively concentrate and aggregate at the surface of the weathered limestones and in tens of thousands of years a soil is created. The **terra rossa** of Mediterranean lands is such a soil. Rather unexpectedly this red soil may be so leached of lime that it becomes infertile and agricultural lime has to be reintroduced by the farmer. **Leaching** is the selective removal of compounds in solution by circulating ground-waters. In the British Isles red soils are almost completely absent despite extensive outcrops of chalk and limestone. This partly reflects the differences in climate between temperate and hot areas but it is also related to their geological history within the last million or so years. Most of the British outcrops during that time have had their original soils modified by glacial action. Many of the present-day soils are thus mixed, having been created in part from glacially deposited materials. Outside but close to the margins of the ice-sheets it is probable that snow cappings persisted for long spells on higher land and this would alter the normal cycle of soil production. This might partly account for the **clay-with-flints** which mantles the crest of the North and South Downs.

In semi-arid regions with a high rate of evaporation soils rich in calcium carbonate may form. The calcium in solution is precipitated to form **caliche** layers, especially during the dry spells. This contrasts with the leaching out of calcite which occurs in more humid areas. Caliche layers are very similar in mode of formation to **iron pans** which consist of hard nodular layers of iron oxides and hydroxides precipitated a few feet below the surface. The prime example of a soil generated by extreme leaching is laterite (p. 40). Large parts of India, Nigeria and Brazil are mantled by this highly aluminous and ferruginous material with its characteristic microspheroidal texture and mottled yellow and red colour. Bauxite is very similar in type and manner of production but it is relatively deficient in iron. This is advantageous to the aluminium industry which works both recent and fossil deposits because the iron is of no value to them and has to be removed.

Transport

Movement of weathered material may be brought about by water, ice and wind. Since these three are closely linked with erosion and deposition in particular types of landscapes they are dealt with later. Gravity is a universal agent of transport, often increasing the effect of the other three agents. Although any movement of material makes erosion possible the effects of gravity and the transformation of gravitational movement into erosional activity are best seen in hilly or mountainous areas. Large talus (scree) slopes may form when angular rock fragments loosened by weathering processes fall down from the rock faces and come to rest at the bottom. On soil- and grass-covered slopes the downwards movement or **creep** is less spectacular, giving rise to a bending over of trees and walls or a curving downhill of rock outcrops (Fig. 37). Soil creep results from particles loosened by alternating wet–dry or hot–cold conditions being pulled downhill by gravity. Freeze–thaw action is also effective and in periglacial areas leads to a more definite flow of soil called **solifluction.** Occasionally the soils are so saturated with water that an observable movement takes place in the form of a **mud-flow.** These are especially spectacular in semi-arid areas after torrential rainfall though they also occur in temperate mountain valleys. When the underlying rocks become saturated with water landslipping takes place. In many instances the movement involves the displacement of thousands of tons of rock and

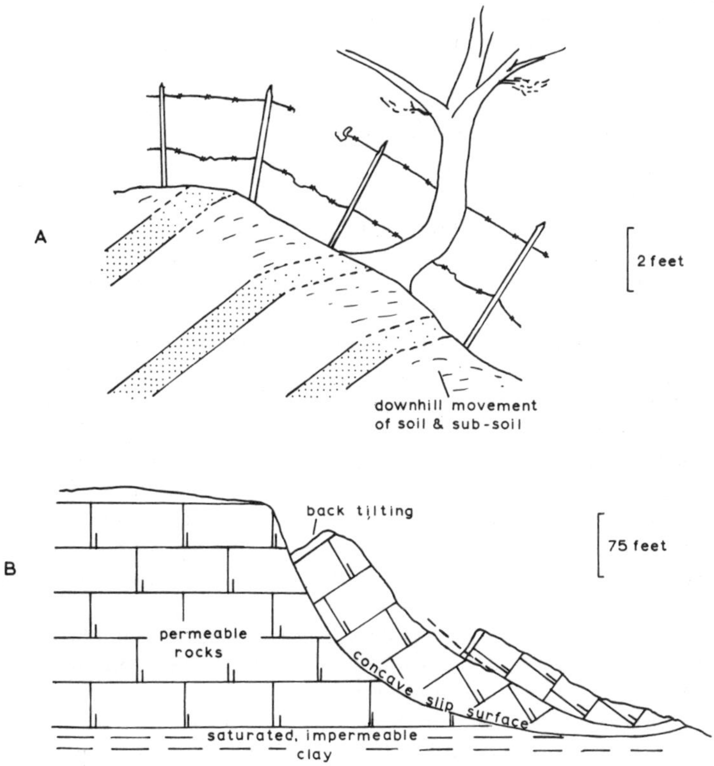

FIG. 37. A: the effect of soil creep. B: landslipping with back tilting of the slipped blocks. The curved slip-planes continue down into the clay

appears to have been triggered-off by earth-tremors. In 1881 at Elm in the Swiss Alps 13 million cubic yards of rock spread over nearly half a square mile of valley floor in just one minute. The debris piled up to a depth of 30 to 60 ft., killing 115 people.

Landslips are not necessarily confined to steep hillsides. Even the shallowest of road or rail cuttings may suddenly be blocked by a slide of saturated material, particularly if the cutting should be made through clays. Cliffed coastal areas are often made unstable by the undercutting action of the sea. **Rock-falls** are caused in this manner. The large-scale collapse of the Chalk cliffs at Folkestone is partly due to constant erosion of the Gault clay outcropping at the base of the cliff greatly aided by ground-water lubrication within the Gault clay. Slipping due to this lubrication means a much greater volume of rock can move; the main planes of movement are concave curves (Fig. 37). This type of landslipping can be prevented by protecting the soft rock at the cliff foot from erosion and by draining away excess ground-water.

Water Erosion and Deposition

The most effective agent of erosion on land-surfaces is moving water. The total volume of water in movement in any one place at any one time is dependent on factors such as climate, ground-water circulation and local topography. In highland areas under subtropical conditions the total amount of run-off is normally much higher than in low-lying temperate areas despite the higher rate of surface evaporation. This is because the average rainfall is higher and the steeper slopes improve what otherwise might be sluggish drainage. Thick vegetation which stabilizes soils can also retard surface and ground-water movements. In flat-lying densely-forested areas the soil lost by erosion may be negligible whereas if bare soil is present there may be a loss amounting to 50 tons or more per acre per year. The deforestation of the farm-lands around the Tennessee River during the late nineteenth and early twentieth centuries led to wholesale erosion of fertile soil and once-prosperous farms rapidly became derelict.

Surface drainage and erosion is the work of rivulets which eventually become streams and finally rivers. The eroding power of water, however, can commence at an earlier stage during a torrential downpour of rain. Soft soils may be actively worn away by the force of the rain at this time. If large boulders are sprinkled throughout them they protect the immediately underlying soil from rain impact and **earth pillars (earth pedestals)** may form. Excellent examples moulded from boulder clays occur in the Innsbruck area of Austria and in the South Tyrol in Italy. The rapid erosion of man-made spoil heaps in industrial areas is often directly related to the gravitational impact of rain and its subsequent flowage down the slopes to a lower level. This is essentially similar to the formation of gullies in the semi-arid 'Badlands' of Dakota. As the water moves over the ground it picks up soil and other particles, so increasing its cutting power. The mobilized particles begin to abrade each other and the solid rocks over which they are being transported. The streamlets thus become more effective as agents of erosion in the process of denudation.

The agents of abrasion carried by streams and rivers form part of the **load.** The physical components are probably the most effective. Some boulders, pebbles and grains are intermittently dragged or bounced along the bottom and form the **traction load.** Smaller grains are borne along within the waters and form the **suspension load.** In times of high rainfall the increased **velocity** and carrying **capacity** may be such that large pebbles are taken up into suspension. Conversely, during drought conditions the coarser fraction of the suspension load may be dragged along the bottom or even deposited. Dissolved within all rivers are a wide range of soluble chemicals which have no abrasion effects. They are the **chemical** or **solution load.** If the waters are draining limestone areas then this load is rich in calcium carbonate; if sandstone-shale areas then it is rich in silica. Although non-abrasive the chemical activity of some streams may be such that chemical solution of rocks may occur. Limestones are particularly susceptible to this kind of attack because they consist of very soluble carbonate minerals.

The effect of the physical load as an abrasive agent depends partly on its speed of impact upon the sides and bottom of the stream. The velocity, in turn, reflects the gradient of the land-surface over which the stream is flowing. The steepest gradients of any river system are found in the headwater regions. Consequently the velocities there tend to be relatively high compared with those near to the river mouth. So it would seem that comparatively active erosion should be taking place at the headwaters. Velocity, however, is not the only factor influencing the abrasive effect of the load; the capacity of the stream, i.e. the amount of material its size allows it to carry, is also important. In fact the individual source streams have not the carrying capacity to bring about extensive erosion. Maximum erosion is found further downstream when the individual streams have coalesced into rivers which have the combined cutting power and capacity of the minor streams though lower velocity. The middle to upper reaches of rivers have a physical load varying between boulders many inches across and sand grains. Velocity as a factor influencing erosion becomes even less important as the river approaches its mouth. In these lowest stages, at heights just above sea-level, erosion is at a minimum. The low gradient and sluggish rate of flow means that only silt and clay particles are transported.

The delicate balance between velocity, capacity and load is

dependent on conditions other than climate or gradient. For example, the flow of water in rivers is rarely smooth. At the bottom and sides frictional forces exert a drag retarding the steady flow. Consequently the physical load may be partially deposited. An extreme though common case of this occurs on the inner side of meanders (p. 83) where a core of newly deposited sediment may be progressively formed. Deposition of silt and clay also takes place in the lowest reaches of rivers particularly during times of low water. At the mouths of rivers where the waters intermingle with more saline sea-waters the clay grains may coagulate and be deposited in such a fashion as to become firmly adherent to the river bottom. This important process helps to create mud- and silt-banks in estuaries and makes them hazardous for shipping. The fine-grained material laid down so prolifically in the lower reaches of rivers during flood conditions is called **alluvium.** The easily eroded **alluvial flats** or **flood-plains** (Fig. 38) adjacent to most major river courses may be very extensive areally, up to 100 miles wide, and very fertile provided the drainage can be improved so as to avoid excessive or permanent saturation.

The larger and more mature a river is then the greater the width of the flood-plain both actually and relatively, relative, that is, to the width of the river itself which only completely fills its valley in the highest reaches. Across the wide flat plain the river follows an extremely irregular serpentine course. The individual loops and curves are called **meanders** (Fig. 38) and may have a radius of up to 5 miles. Towards the inner bank of a meander the water is usually slower running and less deep than towards the outer, and coarser parts of the load are readily deposited, especially during periods of low water. The outer bank, in contrast, is eroded actively owing to the faster and more turbulent currents within the river. Consequently there is a distinct tendency for meanders to migrate across and down the flood-plain by the simultaneous processes of erosion and deposition. In this way the alluvial silts are constantly re-worked. At the same time the erosion in adjacent meanders may be such as to cause them to approach each other closely and the narrow neck of land between may eventually be completely worn away. The course of the river is partly straightened by this and the resultant cut off meanders rapidly become **abandoned meanders** occupied by **meander** or **ox-bow lakes** (Fig. 38).

During times of floods the water may overtop the river banks and inundate the surrounding flat flood-plain and much loss of life may result. Immediately the water flows over the banks its velocity falls and most of the suspension load is dropped. A narrow zone of deposition is created which slowly builds up into a **levée** or raised bank, as in Fig. 38. With larger, more mature rivers such as the Mississippi the levées may attain a height above the flood-plain of 20 ft. or more and a width of 1 to 5 miles. When this stage is reached the dangers of flooding instead of being decreased are, in fact, increased because the level of the

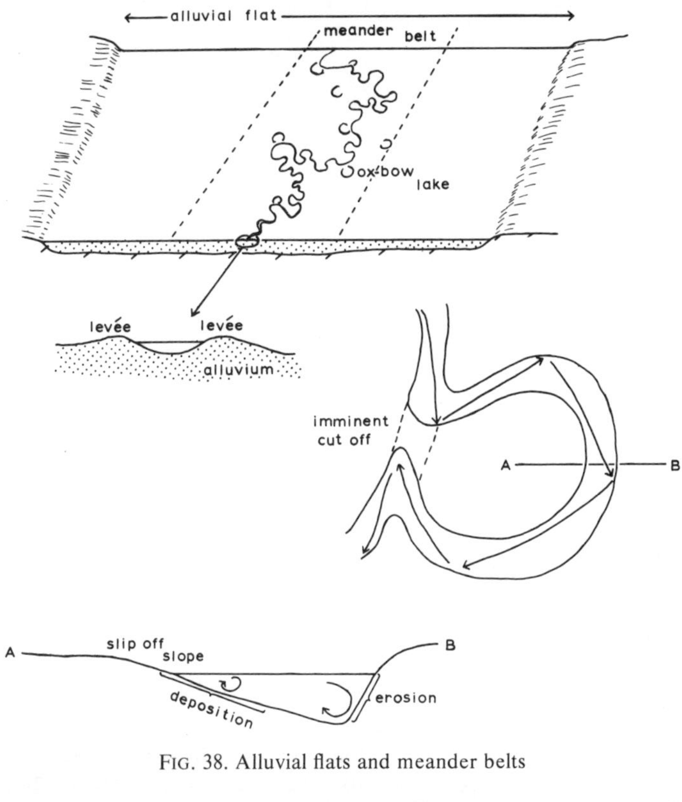

FIG. 38. Alluvial flats and meander belts

water flowing in the river is then higher than the surrounding countryside.

In the middle and upper reaches of rivers the higher gradient and increased velocity of the water cause extensive erosion but prevent extensive meandering. Alluvial flats are much reduced in width and levées are correspondingly smaller in scale. Although intermittent flooding occurs its effects are often less destructive as the river course and flats are more confined by the proximity of surrounding hilly ground.

The headwaters of any river are characterized by the absence of alluvial flats. The land adjacent to the streams is usually steep and the velocity of the streams themselves is high but their capacity and eroding power is low. The valleys are narrow and V-shaped. Their course is also irregular, almost zigzag, as the weaker bands within the country rocks are exploited and enlarged. The inconstancy of direction creates spurs of land which interlock closely.

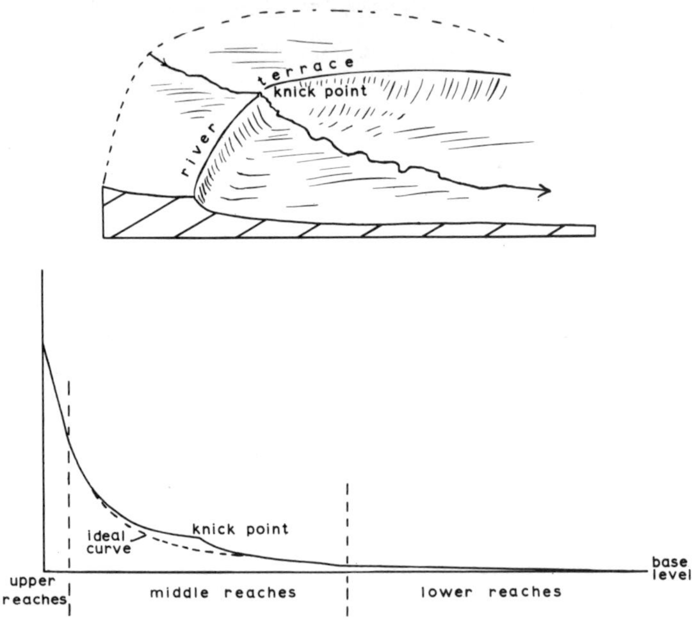

FIG. 39. Long profile of rivers showing the effect of uplift of the land

Our knowledge of the erosional capabilities of a river from its head to mouth helps us to understand the characteristic shape of its **longitudinal profile** (Fig. 39). During undisturbed conditions of formation this becomes a smooth, concave-upwards, asymmetrical curve with a steep fall in the upper and middle reaches decreasing rapidly to a much longer and shallower fall in the lower reaches. If the detailed long profile of any specific river is plotted, however, it is often found that the smooth curve based on theoretical arguments is very rarely attained. There are many reasons for this, such as the lithology and structure of the rocks over which the river is flowing. Hard rocks are more resistant to

PLATE 9

Unconformity. Permo-Triassic conglomerate and sandstone resting on Devonian shales, Thurlestone, South Devon

PLATE 10

Normal fault with downthrow to the right in Lower Carboniferous sandstones and shales, near St. Andrews, Fife

PLATE 11

Overfold resting on top of a thrust plane at Hope's Nose, Torquay. The rocks are Mid-Devonian limestones

erosion than interbedded soft rocks and interrupt the smooth profile. The resultant stepped profile through mixed rock successions becomes even more obvious if the rocks are gently dipping upstream or horizontal because the hard rocks then act as fall makers. The waterfalls produced may be very high. The Niagara Falls are created by the fall of water for 150 ft. over the near-horizontal thick Niagara Limestone which rests on softer shalier beds. Moreover these world-renowned falls, as with many less spectacular ones, tend to be self-perpetuating because the power of the falling water gouges out and undercuts the underlying soft material. The height of the fall is thus maintained possibly for tens of thousands of years. The gouging action of the water commonly creates an overdeepening at the foot of a fall and this is called a **plunge pool.** If, in contrast to near-horizontality, the rocks dip at steep angles the presence of harder bands may be indicated merely by rocky rapids which modify the long profile only slightly.

Probably the greatest effect on profile is caused by changes in sea-level (or **base level**) resulting from either a rise or fall in the water volume of the oceans (**eustatic changes**) or an up- or downwarp of the land. The former may be due to the mass accumulation or mass wasting of ice in circumpolar regions taking place over tens of thousands of years and the latter may be related to slow isostatic adjustments of the crust. Whatever the origin the result is to alter the base level (the limiting factor of downward erosion) of the longitudinal profile. The most interesting results are seen if the sea- or base level falls. The river immediately becomes reinvigorated and recommences downcutting at and near to its mouth towards the new base level. In doing so a new profile is created. The point on the course of the river where the new profile changes into the older is the **knick point** (Fig. 39). The position of a knick point is sometimes indicated by rapids or even a waterfall which need not necessarily coincide with the outcrop of harder beds in the stream or river bottom. The distance upstream of any knick point is variable and depends on how long a period has elapsed since the base level was modified. Also hard rocks delay its migration upstream. Many rivers in Britain have more than one knick point suggesting a number of base level movements.

The lowering of the course of a river means that in time a new flood-plain forms within the old. The old alluvial flats are left high and dry and form paired **river terraces** (Fig. 39) separated from the new plain in the river course only by a knick point. The majority of British rivers have sets of inset river terraces but probably the best known are those associated with the River Thames. The Boyn Hill and Taplow Terraces formed of silts, sands and gravels can be traced upstream as far as Maidenhead at a height varying between 100 and 150 ft. and 50 to 100 ft. respectively. The wide extent of these deposits within the London Basin suggests that the changes of level started in Pleistocene times.

The effect of base level lowering on some meanders is to make the river incise itself into the underlying solid rocks and so create **incised meanders.** If the relative downwards movement of sea-level has been a hundred or so feet then the actual incision may be scenically impressive as with the River Wye at Chepstow and the River Wear at Durham.

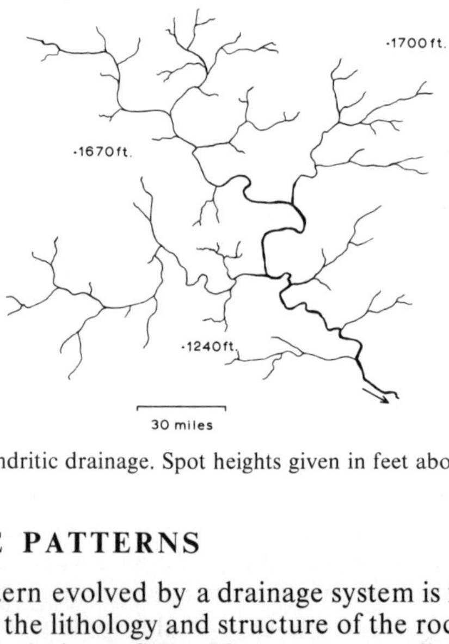

FIG. 40. Dendritic drainage. Spot heights given in feet above sea-level

DRAINAGE PATTERNS

The areal pattern evolved by a drainage system is fundamentally dependent on the lithology and structure of the rocks over which the streams and rivers flow. The more complicated the rock succession and outcrop and the more complicated the folding and faulting then the more complex is the drainage pattern.

Dendritic Drainage

The simplest style of drainage is that developed on a gently and uniformly dipping surface formed of solid rocks of relatively uniform lithology, excluding soluble limestones. The resultant pattern has the appearance of a tree stripped of its leaves and is known as **dendritic** (Fig. 40). The necessary conditions for development are rare in Britain so that no major dendritic system is known here. When seen on a small-scale map in any atlas the Amazon river system has a dendritic appearance. This system, however, is bound to be affected by structure and lithology if viewed in detail.

Trellis Drainage

Trellis patterns are found in the British Isles in the Mesozoic areas of east and south-east England. The basic requisites are gentle fold structures combined with a mixed lithology of hard and soft beds. In the case of the Weald, Cotswolds, Chilterns and Lincolnshire Wolds the resistant beds are various kinds of limestone, chalk and sandstone and they are interbedded with more easily erodible shales and clays. Consequently the hard beds are left upstanding to a great extent by the relatively faster erosion of the softer beds.

The beginnings and development of trellis drainage patterns can be followed if the simplified history of a specific area is considered such as the Weald of south-east England. The rocks in that area were folded into a broad WNW to ESE trending anticlinal upwarp during Tertiary times. On the domed land surface formed of Cretaceous Chalk a series of streams began to flow northwards and southwards into shallow seas. These streams are called **consequents** and they flow nearly parallel to the dip of the strata (Fig. 41). After a few million years the erosive power of the Wealden consequents was such that they cut through the Chalk in the central parts of the Weald into the underlying sands and clays. The lack of lithological uniformity was then rapidly exploited by side-streams working backwards along the strike of the softer beds. These streams are **subsequents** and in time a whole host of them were created. Most flowed along and etched out strike vales in the Weald Clay and Gault clay, leaving the more resistant sandstones and Chalk as low hills. The hills, which include Leith Hill and the Downs, slowly developed their scarp and dip morphology during this stage. Because of the general anticlinal structure (p. 70) the steep scarp slopes face inwards towards the central axial regions of the Weald and the shallower dip slopes outwards. Through the gaps in the hills the trellised river and stream system drained the Weald. In some places **river capture** enlarged some rivers at the expense of others once equally powerful. The original northward-flowing consequent River Medway, for instance, captured the headwaters of the consequent River Darent a few miles to the west. This was accomplished by a powerful subsequent of the Medway eating its way

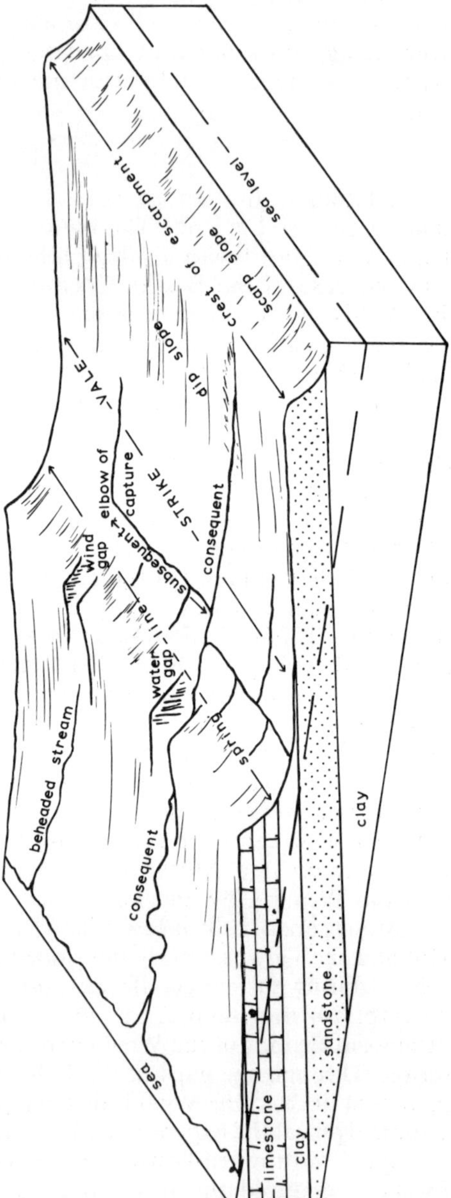

FIG. 41. Trellis drainage

westwards along soft Weald Clay. The point of capture is indicated by a right-angled bend in the river called an **elbow of capture** (Fig. 41). The large gap through the Chalk carved by the original Darent is now occupied by a **misfit** stream of much smaller size. At Redhill and other places the beheading by subsequents was so drastic that gaps are left completely unoccupied by any streams. They are then known as **wind gaps.** A large number of wind gaps cut through the Chalk of the Chiltern Hills. They have evolved in a similar way to those of the Weald. As in the Weald they have been exploited by man through the ages in improving communications and at present many important roads and railways pass through them.

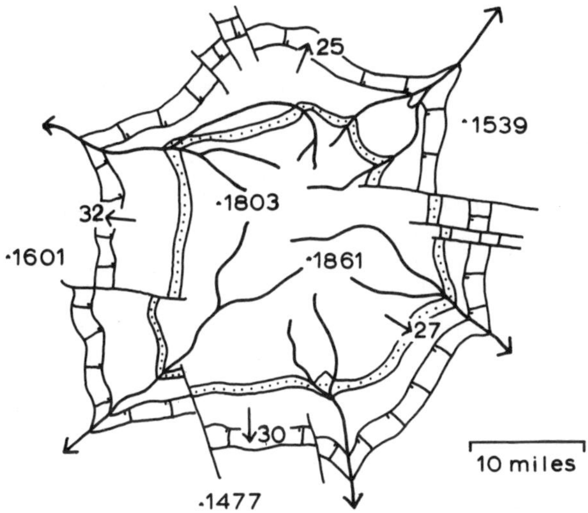

FIG. 42. Radial drainage

Radial Drainage.

This pattern develops on a rising domed surface of beds of uniform lithology. The streams and rivers created on the uplifted surface drain outwards as if from the hub of a wheel (Fig. 42). Examples of simple radial drainage can be observed on man-made spoil heaps especially in mining areas but natural British examples are very rare. The best occurrences abroad are around the flanks of volcanic cones such as Fujiyama in Japan or Vesuvius in Italy where the soft pyroclastic debris is quickly etched by streams. The Lake District in north-west England is commonly quoted as illustrating radial drainage as indeed it does. The origin, however, is far from being a simple matter and needs linking up with a process called **superimposition.**

Superimposed Drainage

Superimposition occurs when a particular and often mature system of rivers created on an uplifted surface cuts down through that surface into underlying older rocks with a different structure. The Lake District is a prime example of radial drainage superimposed from a younger dome-shaped cover on to Lower Palaeozoic rocks. The older rocks are not domed in structure but are arranged anticlinally about a WSW to ENE axis. The time when the main superimposition occurred is not precisely known but it was almost certainly during Tertiary times. Even more doubtful is the lithological nature of the domed surface uplifted at the time. Most geologists favour a Cretaceous Chalk cover even though the Chalk has now been completely eroded away.

Many other rivers in Britain show evidence of superimposition along part of their present course. The River Derwent at Matlock flows southwards through a narrow limestone gorge instead of continuing along its normal path over soft shales which are only a few hundred feet to the east. Likewise, the River Avon at Bristol cuts a spectacular gorge through Carboniferous limestones instead of following a much easier route along softer Upper Carboniferous shales.

Antecedent Drainage

This is a localized effect caused by a slow upwards movement of the land during and after the establishment of the main drainage pattern on its surface. It is always difficult to detect though less so in mountainous headwater areas than in lowland areas. This is because the uplift in mountains brings about an exaggeration in the normal V-shaped form of the valleys. The V-shape becomes much more narrow and there is progressively produced a very deep series of valleys sometimes with a vertical incision of many thousands of feet. The headwaters of the Indus and Ganges rivers in the Himalayas are characterized by these abnormal valley depths and it has been suggested that the cause is erosion keeping pace with the slow uplift of the mountain range. As Britain has been a relatively stable area since the establishment of its main drainage patterns there is no definite evidence of antecedent river courses.

Peneplains and Monadnocks

It is very rarely that a river system becomes so permanently established that erosion of the land continues until no upstanding areas remain. Before this stage is reached the base level is usually either altered by eustatic movements of sea-level or isostatic warping of the land. A common result of these movements which may well be simultaneous is to raise the land-surface and reactivate the rivers. Consequently, the original upstanding and dissected topography is maintained. On the rarer occasions of prolonged **stillstand,** that is with sea-level remaining constant, river erosion proceeds to its logical end and produces an areally extensive and relatively flat low-level plain embracing the whole river system. This is a **peneplain.** The rocks, whatever their original structure and lithology, are all worn down although more resistant rocks may remain for a long time as isolated **monadnocks** standing well above the level of the surrounding plain.

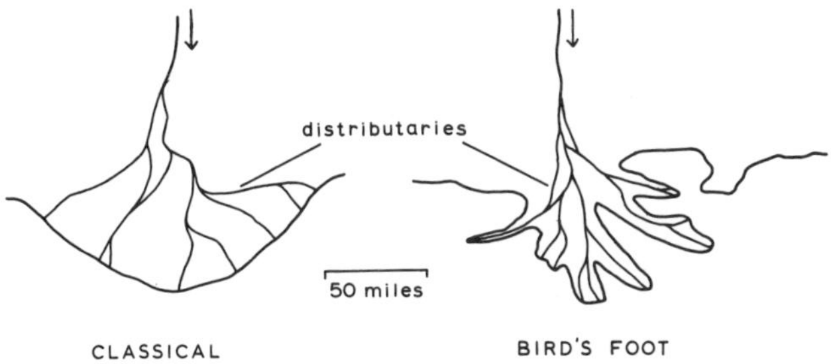

FIG. 43. Deltas

Deltas and Estuaries

All river systems tend to lose their distinctive inland drainage pattern as they approach their place of disgorgement into the sea. The main trunk rivers coalesce into a much larger whole which meanders extensively before depositing its load. The actual point of entry and maximum deposition into the sea may either be in the form of a delta or estuary.

Deltas are dependent for their formation on a rapid large-scale deposition of material under general conditions of low current and wave activity. The best-known deltas are associated with large rivers disgorging into slightly protected seas. The Nile delta of approximately 50,000 sq. miles area pours its material into the south-eastern end of the non-tidal Mediterranean Sea. The gradient is so low at its mouth that the main trunk feeder splits up into innumerable **distributaries** which splay out and deposit fluviatile material in a fan-shaped fashion (Fig. 43). This creates the familiar △-shape. The deltas of the Ganges, Niger and Rhône rivers are very similar.

The delta of the Mississippi River has a modified pattern partly due to subsidence of the mouths keeping pace with sedimentation and partly due to more powerful longshore and off-shore currents. The distributaries finger out into the Gulf of Mexico and deposit material in such a fashion that they have the appearance of a bird's foot. Hence the name **bird's foot delta** (Fig. 43).

Fine-grained sands, silts and muds are the typical deposits of all mature deltas. In the Lake District, however, there are small lakeside deltas formed of much coarser sands and pebbles. The reason for this is that many of the side-streams pouring material

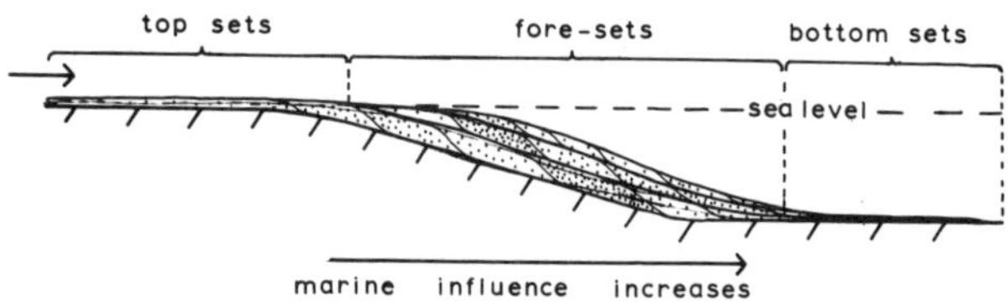

FIG. 44. Cross-section of a delta

into the lakes have a very steep gradient and high velocity. In addition it is probable that maximum deposition occurred in late Pleistocene times when the British climate was much wetter than at present.

Both large and small deltas have a distinctive cross-section from head to foot which can be divided into three parts, based on slight variations in the character of the sedimentation (Fig. 44). In major deltas margining seas the distributary area is characterized by finely laminated non-marine muds and silts. These are **top-set** deposits and are laid down nearly horizontally both within the active rivers and in the lagoonal and swampy areas between the rivers. At and just off the mouths of the distributaries the sediments are coarser fluvial silts laid down on shallow slopes of a few degrees inclination. These **fore-set** deposits progressively build outwards the submarine foundation of the delta and eventually top-set sediments raise the level up to or just above sea-level. In this way deltas encroach into seas and oceans. Away from the subaerial part of the delta in the **bottom-set** areas finely laminated near-horizontal fine silts and muds are laid down. They are mixed in nature, some being fluvial in origin, others having a more marine aspect with sporadic marine organic traces. The influence of fluvial sedimentation decreases rapidly at these outer extremities of the delta and is replaced by normal marine deposition.

Estuaries are funnel-shaped indentations of the coast which pass headwards into a minor or major river system. The Severn estuary has almost an ideal form whereas those of the Mersey, Humber and Thames are a little more irregular. Some estuaries may be quite long. The Gironde in south-west France is nearly 60 miles long although its width is never greater than 6 miles. Because of their shape rising tidal currents become constricted as they move towards the head and pile up giving much higher tidal ranges than in the more open adjacent seas. These same currents also have a scouring action, especially during ebb-tides, and this keeps the deeper channels clear of excessive silt and mud. The chief deposits are fluvial laminated silts and muds some of which form banks off-shore and others mud-flats around the margins. The very extensive nature of the latter is revealed at low tides. For example, at the mouth of the River Thames over a mile of flat is exposed daily. The topography is flattish but is often disturbed by live and dead mussel banks, apart from minor drainage channels.

The tendency of all estuaries is to silt up, and this process is commonly expressed in the progressive colonization of the higher parts of the mud-flats by plants and animals. Marsh or swampy ground is created and eventually this builds up to low-lying but less saturated ground. At the head of the estuary the marshes are freshwater. Further downstream because of the influence of the more saline sea-waters the marshes are more brackish in type.

Seas and Oceans

The ultimate destination of all river-transported material is the sea. There the load is eventually deposited and lithified into solid rock. The fact that marine sediments are protected against the ravages of subaerial attack means that they have a greater chance of survival than terrestrial deposited material. It is not surprising, therefore, that most sedimentary rocks preserved in old successions are marine in origin. It should not be assumed that materials come to rest on the sea-bottom and immediately become lithified. The powerful tides and currents make certain that this is not so. The tidal influences are of paramount importance along coasts. The muds, sands and gravels which fringe the shore-line are in constant movement, particularly during periods of spring tides or storm surges. The rate of movement of material away from the shore-line is often enhanced by submarine canyons, the heads of which may occur only 20 or 30 ft. below normal low-water mark or just off-shore. The canyon heads usually have a dendritic pattern and steepen rapidly before entering into the flatter main trunk canyon so any sedimentary load quickly slips down them. The slipping movement may be in the form of a turbidity current (p. 30) and successive currents may carry a particular load in a bouncing fashion further down the main canyon away from the coast. Submarine canyons indenting the continental shelves around the Pacific and Atlantic Oceans often extend for tens of miles out from the coast and have depths up to many hundreds of feet. The best-studied examples are off the Californian coast near Los Angeles.

The edge of the continental shelf is usually taken as the point where the average depth exceeds 100 fathoms and is marked by an overall rapid increase in gradient of the sea-bottom. Little is known about deposition within the deeper parts of the oceans away from the shelves. Echo-soundings show that the bottom topography is much more variable than originally envisaged. A very pronounced ridge, for instance, extends from Iceland to Antarctica in the mid-Atlantic. Sedimentation, therefore, is probably highly irregular with no deposition in some places and thicker deposition in others. In some deep basins around the Pacific margins which descend to depths of 4,000 ft. or more very fine-grained dark muds are deposited. On the interbasinal ridges at depths of about 1,000 ft. the sediments are clean-washed sands and calcareous silts often rich in calcareous and siliceous organic debris. In other places the power of the deep-sea currents may be such that only bare rock is present. In quieter areas manganese oxide nodules up to 3 ft. in diameter may be precipitated.

Well away from shelf seas the dominant type of sediment is fine-grained organic ooze. The organic content is usually high because very little coarse land-derived detritus ever reaches much beyond the margins of the continental shelf. There is no dilution, therefore, of the dead organic concentrations. The distribution of organic oozes on ocean bottoms is inferred more from the nature of life and processes at the surface than from factual bottom data. Deep-sea coring is becoming more frequent now but large areas still need studying before an accurate picture of oceanic sedimentation can be deduced. On the basis of our knowledge of surface life at present several general points can be made. In warm currents calcareous foraminifera and siliceous radiolaria may be prolific. On death their tests rain down slowly on to the ocean bottoms and form oozes. The distribution of the oozes depends on the bottom currents and the fact that calcareous skeletons are more soluble at depth than siliceous ones. Hence the ooze actually formed on the bottom may be quite different from what the surface conditions indicate.

A most intriguing sediment found in the greatest depths is clay, some of which is red in colour when fresh but most of which tends to be grey. Within it there is fish and other organic debris sometimes intermixed with degraded glassy materials of igneous nature. It has been suggested, therefore, that some of the clay may have originated from the breakdown of volcanic dust carried by the earth's wind system well away from the parent land areas.

Shore-lines

The beaches and immediately adjacent land and sea have always been of much interest to geologists because of their general accessibility and clues they give as to the nature of marine erosion and deposition. The attainment of information is often made more exciting by the constantly changing appearance of the coastlines. The cliff-foot sand and pebble beach of the summer season may be completely washed away during the winter months by stormy seas. Simultaneously cliff falls may modify the shape and size of the cliffs.

The persistent attack of waves on land is visibly expressed in many ways apart from cliff falls and slips. In dipping strata hard rock bands tend to persist as headlands whereas the softer interbedded rocks are etched out quickly. The relatively hard Chalk at the Needles or Flamborough Head and the dolerite sill at the northern end of the Firth of Forth road and rail bridges form very pronounced headlands. The presence of headlands and associated embayments in turn has a marked effect on the crests of incoming waves. This effect is known as **wave refraction** and is caused by the wave crests (or fronts) being retarded and concentrated where they meet the headlands. In the centre of the bays the wave crests move at a faster rate and the wave pattern produced tends, as a result, to parallel the shape of the coastline. The most important consequence of this is the concentration of wave energy and eroding power on the harder rocks of the headlands (Fig. 45).

Wave erosion is accomplished by hydraulic pressure, abrasion by pebbles and, to a lesser degree, by chemical solution. During storm conditions the pressures exerted by breaking waves commonly reach 3 tons per sq. foot. Man-made harbour and sea walls can rarely withstand prolonged pounding under these conditions. The disastrous results are often made worse by the presence of drainage pipes or fissures containing air, because the air may be compressed to such a high degree that it explosively loosens adjacent stonework and foundations. The same is true for hard, well-jointed and bedded rocks at headlands. The waves and attendant air pressures enlarge and wedge open the natural fissures to such an extent that **blow holes, sea caves, arches** and **stacks** eventually result (Fig. 46). Blow holes are enlarged vertical fissures of pipe-like form through which sea-water spouts

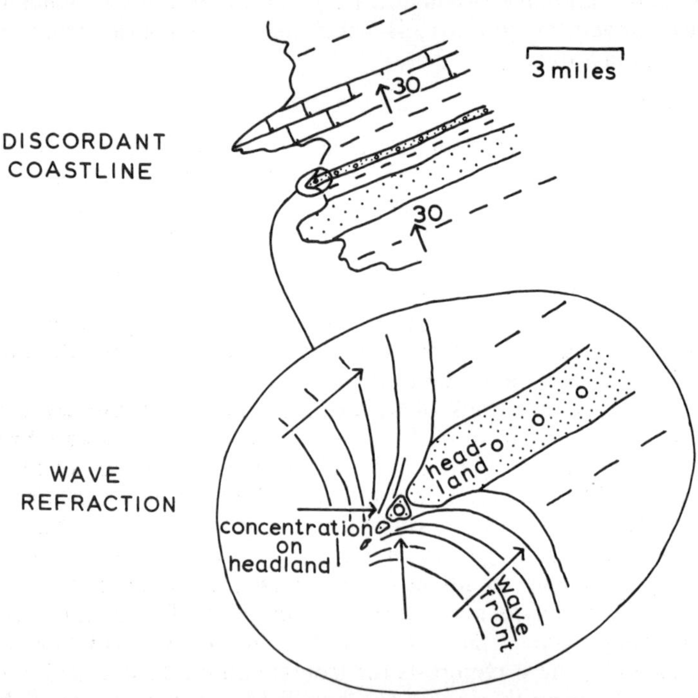

FIG. 45. Discordant coastlines and wave refraction

during high tides and more especially during storm conditions. Sometimes two fissures enlarge into caves on opposite sides of a hard rock rib. As the caves progressively enlarge there comes a time when they coalesce and a natural arch forms. Ultimately the arch itself collapses leaving behind an isolated stack. The south coast of Britain is particularly well endowed with examples of structures, for instance sea caves near Portland Bill, sea arches at Stair Hole and Durdle Door in Dorset. Old Harry (Chalk) at Studland, the Needles (Chalk) in the Isle of Wight and the Old Man of Hoy (Old Red Sandstone) in Scotland are some of the best known sea stacks. Good examples of blow holes occur at Flamborough Head.

The concentration of wave attack on headlands resulting in their progressive physical decay creates eventually a **wave-cut platform** (Fig. 48). This is a near horizontal surface within the zone of breakers which cuts across all the beds whether they be dipping or not. At low tide on cliffed coasts the bare rocks forming the platform may be nearly completely exposed together with a **wave-cut notch** (Fig. 48) at the foot of the cliffs. On non-cliffed coastlines the platform may be covered by a thin mantle of gravels, sands and muds and the notch obscured by slumped material. Often the detection of the platform and notch is obscured by pebbles and sand which have been moved along the

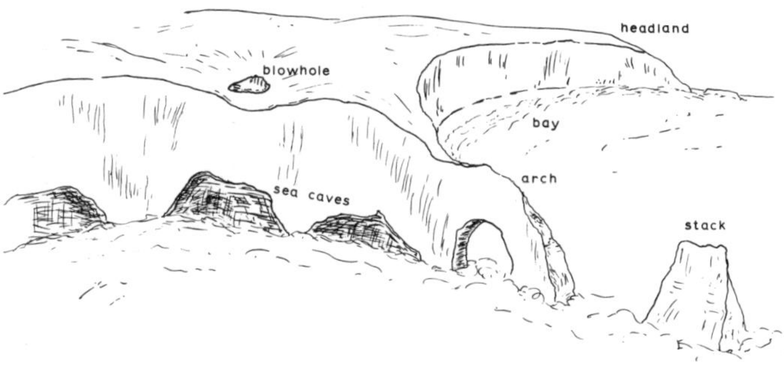

FIG. 46. Arches, stacks, sea caves and blow holes

beach by longshore currents. These beaches appear to be permanent but this is deceptive because they are being fed constantly with new materials from further along the coast. Chesil Beach which extends from Bridport in Dorset to the northern tip of the Isle of Portland, a distance of 19 miles, is probably of this kind. If the supply of material is stopped at source as it might well be by some man-made pier or jetty then the beach may be 'starved' and disappear revealing the underlying platform and notch. In other words drifted beaches often have a protective influence against the eroding power of waves.

The lengthwise movement of materials along shore-lines is called **longshore drift.** The important factors determining the direction of longshore drift are the waves and currents which in turn are controlled by the prevailing winds and the **fetch,** or uninterrupted distance over which they have blown. If the on-shore winds and waves impinge at right-angles to the shore then little or no drift occurs. An oblique set of winds and waves, however, immediately produce an intermittent zigzag motion of the particles with a net forward movement which may amount to a few inches or many feet over a short period of time (Fig. 47). The effect can be envisaged in terms of a **swash** movement of the breaking waves and transportation of particles obliquely up the beach slope followed by a **backwash** movement straight down

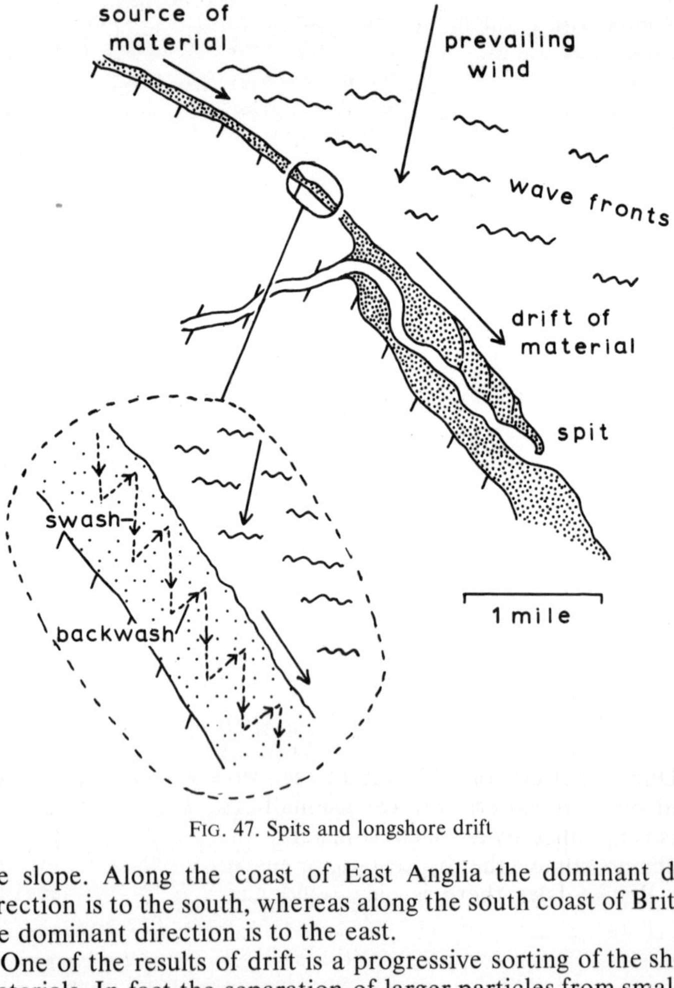

FIG. 47. Spits and longshore drift

the slope. Along the coast of East Anglia the dominant drift direction is to the south, whereas along the south coast of Britain the dominant direction is to the east.

One of the results of drift is a progressive sorting of the shore materials. In fact the separation of larger particles from smaller, as where pebbly beaches grade into sand, is often the only clue that longshore drift processes have been at work. For example, Chesil Beach changes eastwards from a coarse sand at Bridport into a pebbly gravel at the Isle of Portland, the drift being from north-west to south-east.

A more positive clue to longshore drift is given by **spits** (Fig. 47). In effect these are prolongations of beaches from the shore-line into the sea. They tend to occur where the coast is indented by estuaries and the suggestion has been made that formation is directly influenced by the meeting of the river and longshore currents. The disruption of the longshore drift pattern results in the deposition of the materials being transported. The spits formed in this fashion commonly divert the mouths of rivers many miles as with Orford Ness and the River Alde at Orford. Others just restrict access such as at Exmouth and Spurn Head on the Humber. The importance of current variation and conflicting current directions on sedimentation adjacent to coasts is often seen at low tide or detected by taking soundings of the sea-bottom. There then may be observed or deduced **sand-bars.**

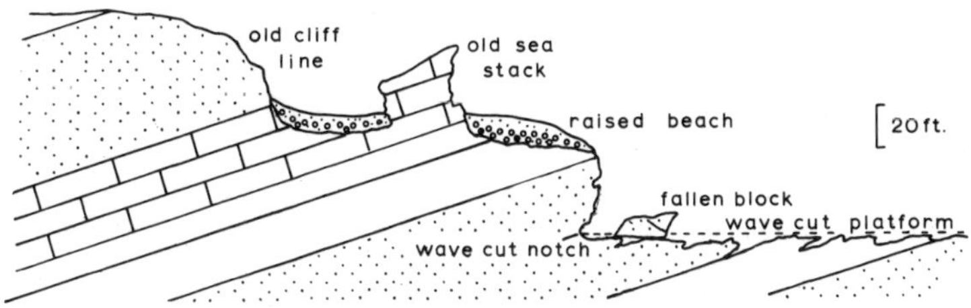

FIG. 48. Coastal erosion and the effect of uplift of the land. The old cliff line is progressively degraded by normal subaerial processes

Some of these are probably old spits which have become submerged by current change and consequent lack of sediment supply. Others may be created independently by current action well off-shore. Many of the Niger distributaries in West Africa have submerged sand-bars running nearly completely across their mouths creating hazards for shipping.

During periods of stillstand the sea slowly encroaches on the land by marginal erosion. Occasionally the steady abrasive process is modified by a relatively faster displacement of the position of the coastline either by isostatic or eustatic movements. Around the British Isles there is clear evidence that the sea-level has fallen about 25 ft. since late Pleistocene times. The most obvious feature is a flat platform of variable width, as around the Firth of Forth, on which rests a veneer of pebbles and sand intermingled with marine fossils. This is a **raised beach** (Fig. 48). Usually the materials forming the beach are loose and uncemented with a covering of soil but in some limestone areas, such as the

Gower in south Wales, the beach is hard-cemented by calcite derived from the underlying rocks. On the landward side there are sometimes seen old shore-line features like cliffs, sea caves and stacks. One of the best-known old cliff lines in southern England occurs on the North Downs extending westwards from west Kent. A distinct notch backed by a low degraded cliff can be observed on the northwards-facing dip slope of the Chalk at an altitude varying between 550 and 600 ft. above sea-level. It appears to have been cut by an early Pleistocene sea existing on the present site of the London Basin. In south Wales, Cornwall, Devon and other hilly parts of Britain the remnants of raised marine platforms have been detected ranging in age from late Tertiary to Pleistocene. Their heights above sea-level commonly range between 25 and 300 ft. but others have been inferred at even higher levels, e.g. the 430 ft. platform of Cornwall. The land has clearly been moving upwards relative to sea-level for several millions of years in these areas.

Coastal Patterns

The relative movements of land and sea play an important part in the development of any shore-line. The actual coastal pattern which evolves is more closely related, however, to the structure and lithology of the rocks being eroded by the sea. For instance, along a **concordant** coastline where the rocks strike parallel to the coast and are uniform in lithology then the coastline will recede at a uniform rate only partly modified by differential erosion along joints, faults and bedding planes. The process of differential erosion along a concordant coastline where the rocks differ in lithology is well illustrated in south Dorset (Fig. 49). There the almost vertical Upper Jurassic limestones have been breached by the sea and softer Cretaceous sands and clays behind systematically worn away to form coves and bays. Lulworth Cove is typical. A similar effect can be seen down the eastern side of the Pacific Ocean and the term **Pacific-type** coastline can be used in place of concordant coastline. If adjacent soft rock bays link up laterally and create a large number of off-shore hard rock ribs and islands as on the Dalmatian coast of the Adriatic Sea, the pattern can be called **Dalmatian-type.**

Discordant coastlines, the antithesis of concordant, are observed where the rocks strike obliquely or at right-angles into the sea. Differential erosion in a mixed group of rocks produces bays separated by headlands of the harder beds (Fig. 45). The coast at Swanage in east Dorset and in the south-west of Ireland are perfect examples. These coastlines are also termed **Atlantic-type.**

In parts of southern Britain the various styles of coastline and associated estuaries have been exaggerated and modified by upward movements of sea-level, probably in Recent times. The lower reaches of the original rivers and alluvial flats have, in consequence, been drowned to various depths for many miles upstream and the rivers at their head have had to establish a new long profile. This **coastline of submergence** is known as the **ria type.** As the drowning appears to have been accomplishedquickly, the land adjacent to the enlarged estuaries commonly slopes steeply towards them and may even be cliffed owing to subsequent wave erosion. A good example of a ria is Milford Haven in south Wales.

The coastlines in areas which have been extensively glaciated are often characterized by very deep steep-sided waterways extending many miles inland. The original estuaries around the

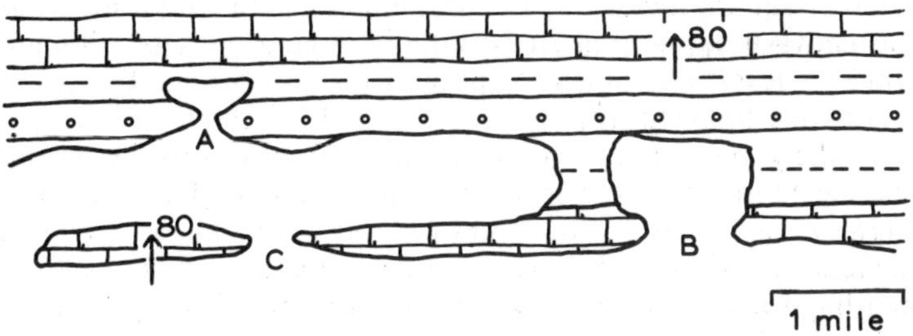

FIG. 49. Concordant coastlines. A: initial breaching of hard beds. B: exploitation and enlargement by erosion of softer beds. C: late stage with complete isolation of hard rock ribs (islands)

coasts of Norway and Scotland were all overdeepened during Pleistocene times to produce complex systems of **fiords.** Some of the very steep-sided Norwegian fiords extend for more than 50 miles inland and are many hundreds of feet deep. The eroding valley glaciers seem to have gouged out the soft and hard rocks along the course of the original valley, steepened the sides and straightened the course. In many cases this process was eased by the presence of large fault zones which traverse the estuaries and valleys from end to end. A common feature of the mouths of most fiords is a **threshold** or zone of shallows (about 20 ft. deep) created by a solid rock bar with or without a mantle of moraine.

Glacial Erosion and Deposition

At least one tenth of the world's surface is covered by ice at the present day. There are ice-sheets in polar and high latitude circumpolar regions and at lower more temperate latitudes ice cappings are produced as a result of increased altitude. There are few major peaks which do not have a permanent or semi-permanent snow and ice cover. The peak of Mount Kilimanjaro (19,320 ft.) in East Africa, situated almost on the Equator, is rarely free from ice. In low latitudes the glaciation is invariably expressed in the form of relatively small ice cappings and emergent glaciers which descend into the valleys; circumpolar ice-sheets are regional in scale. The area influenced by ice accumulation and degradation is much greater of course than just the area covered by ice, and many hundreds of miles away from the nearest ice-sheet the ground may be permanently or semi-permanently frozen to a depth of tens of feet. This is a state of **permafrost.** Meltwaters from ice-sheets may transport materials carried within the original ice for many miles before depositing them. Deposition of this kind is **fluvioglacial** in nature and many thousands of square miles may be affected adjacent to the ice front. The **outwash apron** created is initially devoid of vegetation and hence is desert-like and susceptible to wind erosion. The strong cold winds blowing from the ice cap pick up the finest silts and carry them outwards, and sometimes even upwards, before depositing them as a mantle of structureless **loess.** Loess blankets extensive areas in North America and Asia. There are small pockets in Kent and Essex. Large ice-sheets such as those mantling Greenland and Antarctica at present also edge out into the sea and break off to form **icebergs.** The load of debris carried within them may be deposited eventually many hundreds of miles away from the source.

The causes of extensive ice formation as seen in the last great ice-age, the Pleistocene, are unknown, though complicated hypotheses have been suggested by cosmologists. All the geologist sees is the evidence of glaciation and the more recent the glaciation the stronger is the evidence. A large amount of information has accumulated about recent ice action and the last major phase of glaciation about half a million years ago. Less is known about the Permo-Carboniferous ice-sheets which affected the southern continents and parts of India about 290 million years ago and even less about Pre-Cambrian ice-ages of which there have been several.

Highland Glaciation

During very cold and wet climatic conditions snow and ice accumulate on high ground. The snow which settles or avalanches into hollows progressively converts into ice by compaction and the ice, in turn, welds to the underlying rock, especially along crevices. In the warmer season partial melting tends to occur at the base of the ice patches and the gravitational pull carries the ice plus wedged-off rock fragments down slope. In this way the base of the slowly moving ice begins to act in the form of an abrasive rasp gouging out the country rocks. Eventually large crescentic or horseshoe-shaped hollows are created with steep backs and shallow lips. They are known as **corries, cwms,** or **cirques** (Fig. 50) and are, in effect, ice accumulation grounds for the valley glaciers. The fact that the rock load in ice is capable of abrasion means that corries can enlarge themselves by carving backwards into the hillsides. In doing so they are greatly assisted by freeze-thaw action on the solid rocks exposed in the large crevasse or **bergschrund** on the steep headward side. The recession of two adjacent corries under this combined attack produces a knife-edged ridge or **arête** (Fig. 50). Striding Edge in the Lake District is a good example.

At the head of any pre-existing mountain valley several corries may be created all feeding ice into it. Where the ice converges the scouring and plucking action at its bottom and sides are intensified many times forming a sudden steepening in the long profile—the **trough end** (Fig. 50). As the ice fills and moves along the valley what was once a normal V-shaped transverse and smooth longitudinal profile becomes slowly modified into a U-shaped and stepped profile (Fig. 50). The U-shape results from the greater abrasive power of ice as compared with water. The ice can hold its abrasive 'tools' more firmly and the abrasion takes place across the whole of the lower surface of the ice and at both sides. The U-shape is emphasized by the truncation of spurs and straightening of the valley course. This is related to the less flexible nature of moving ice. The stepped longitudinal profile may be formed in at least two ways. Where minor tributary glaciers join the main trunk glacier the eroding power may be increased sufficiently to abrade the rocks more and form a step. On the other hand the same effect may be produced by differences in the hardness of the rocks traversed by the ice, the harder rocks resisting erosion and standing out as steps. Bottom irregularities of this kind are usually reflected upwards through the ice and are expressed at the surface by ice falls and extensive crevassing. After the ice has melted any basins which may have been carved out of the solid rock of the valley floor are filled with water and become lakes. Some of these, such as Lake Ullswater, are long, narrow and deep and are called **ribbon lakes.** The differential eroding power of glaciers is also confirmed in once-glaciated valleys by the presence of **roches moutonnées**

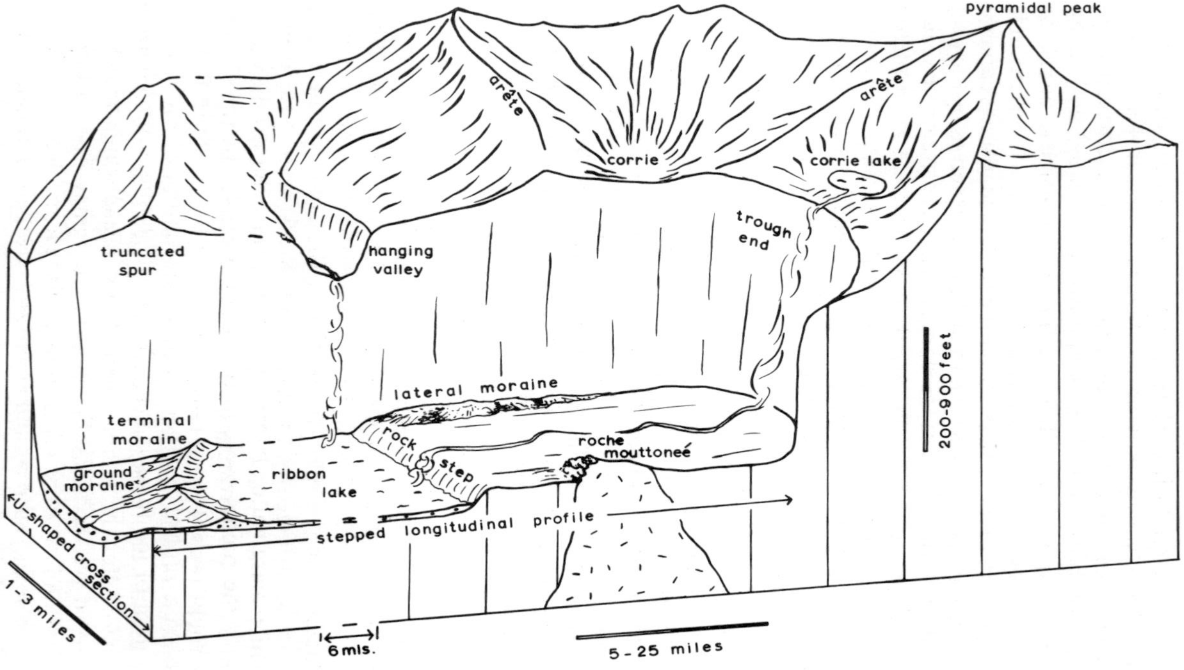

FIG. 50. Highland glaciation. Ribbon lakes are long and narrow being confined by the valley sides and a barrier (e.g. terminal moraine) towards the mouth of the valley

(Fig. 50) which are upstanding isolated rock masses. They are asymmetrical mounds of country rock with a smooth but striated surface on the upstream side and a very jagged, plucked steeper face on the downstream side. **Glacial striae** are often found on ice-smoothed rock faces. They are shallow grooves and scratches caused by the abrasive rock load in the moving ice and a study of them can indicate local directions of ice movement.

The movement of glaciers down valleys during prolonged cold phases is very slow and may only amount to a few feet in a year at the front end or **snout.** The slow advance may be retarded even more if a warm climatic phase intervenes and prevents ice accumulation in the feeder grounds. Any lengthening of the warm phase and the glacier begins to melt and, in effect, recedes headwards. Most of the world's glaciers have intermittently receded and advanced over the last thousand years or so. In 1961 some of the Swiss glaciers were proved to be receding at the somewhat abnormal rate of 100 ft. per year.

Perhaps the most spectacular evidence of glacier melting (**ablation**) and recession in highland valleys are irregular mounds of **moraine,** that is rock load dropped by the ice. At the snout **terminal morainic** debris may be deposited in sufficient quantity to form a barrier across the valley (Fig. 50). Behind the barrier meltwaters may then accumulate to form a lake. The U-shape is modified under these circumstances by lacustrine sedimentation and an alluvial flat-bottomed shape created. The spasmodic retreat of a valley glacier is commonly indicated by a series of terminal moraines and old lake sites within any particular valley. The term **recessional moraines** can then be used in place of terminal moraines. The material forming the moraines is boulder clay, just as it is in the case of the **lateral** and **ground moraines.** Lateral moraine is laid down at the margin of the ice and forms boulder-strewn slopes and ground moraine is deposited from the bottom of the glacier. In large valleys with many tributary glaciers the marginal or lateral moraines may become **medial** in position at the points of confluence. On melting, this centrally situated material is deposited to form either part of the terminal moraine or a central boulder train extending up the valley.

One effect of ice thaw is the large-scale dumping of boulders; some of these may be many miles away from their original outcrop and are then known as **erratics.** Ice thaw also generates streams; the meltwaters flow on and within the ice gathering morainic debris and eventually emerge at the margins or at the snout of the glacier. Part of the load is then dropped but most is transported for many miles depending on local topographic and gradient conditions. The materials forming the load are said to be fluvioglacial in nature.

If the glacial conditions affecting a mountainous area progressively ameliorate over many thousands of years as they have done in Europe since Pleistocene times then there comes a time when very few glaciers occupy the valleys. The characteristic U-shape of fiords in Norway and many valleys in Snowdonia, the Southern Uplands and Scottish Highlands, together with spreads of morainic and fluvioglacial debris, indicate the presence of glaciers until near-Recent times. In addition, the smoothed sides of the major valleys and fiords are commonly indented by minor U-shaped tributary valleys which hang many hundreds of feet above them. The form and position of these **hanging valleys** (Fig. 50) is related to their original occupation by smaller glaciers of lower eroding power than that of the main glacier. The small streams which plunge over the lip create spectacular waterfalls. Corries may be occupied by small lakes (**tarns**); these may overflow over the lip and occasionally waterfalls form.

Lowland Glaciation

During times of extreme and persistent ice formation glaciers and ice cappings may extend outwards into low ground well away from their highland accumulating areas. The glaciers coalesce into extensive ice-sheets which advance across low-lying ground smoothing and abrading as they do so. In Pleistocene times ice-sheets advanced many times across Britain reaching as far south as north London, the northern Cotswolds and south Wales (Fig. 51). Over all the lowland to the north evidence of glaciation is abundant. In the Midlands the boulder clays contain erratics derived from the Welsh and Lake District mountains as well as local materials so there is little doubt as to the source of the ice. Northern England has a scatter of erratics derived from the Southern Uplands and Scottish Highlands and in East Anglia there are boulders which could only have been transported from Scandinavia. Because of topographical irregularities and several phases of advance and retreat spread over several hundreds of thousands of years the boulder clay mantle in Britain varies considerably in thickness from one place to another even over short distances. The greatest thicknesses (tens of feet) are found where old valleys have been infilled (or plugged) and on the lee side of upstanding rock masses. The basalt-infilled vent on which Edinburgh Castle stands or the trachyte plugs of North Berwick and Traprain Law in East Lothian each have a wedge of glacial debris on their eastern side. The inference is that ice movement was from west to east in these areas. On the upstream side of masses of this kind steep craggy slopes with abundant striations occur, hence the name **crag and tail** (Fig. 52) is applied to the structure as a whole.

The boulder clay deposits created by ice-sheets are morainic in origin. The materials transported by the ice, however, have a more obscure pattern than those of valley glaciers, for example it is impossible to distinguish lateral and medial moraines. Terminal moraines are more easy to detect. At Escrick, about 10 miles

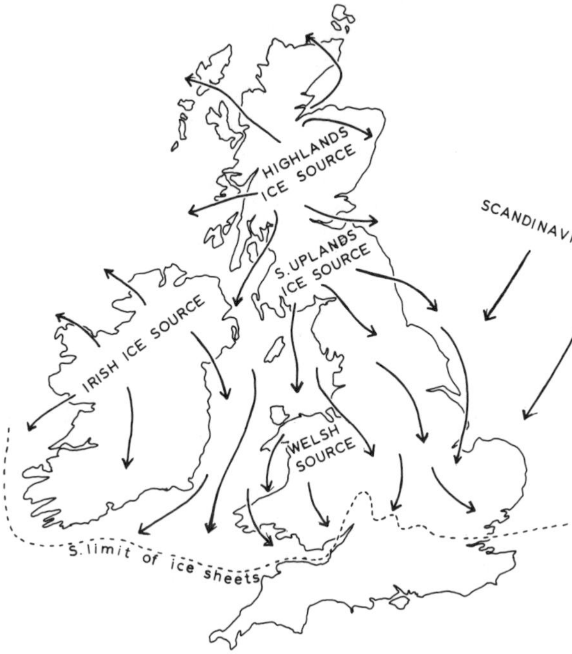

FIG. 51. Glaciation of the British Isles

to the south of York, a sand and gravel ridge which runs roughly east–west for several miles across the flat Vale of York is probably a terminal moraine of Pleistocene times. Peculiar depositional patterns of moraine can be seen in areas such as the Solway Plain, Midland Valley of Scotland and Antrim. There are found in great abundance low elongate mounds or **drumlins** (Fig. 52) formed nearly entirely of boulder clay. The surface effects are very distinctive and are commonly referred to as 'basket of eggs' topography. A typical size for a drumlin is about 800 yds. long by 250 to 500 yds. wide with heights of 30 to 50 ft. An important

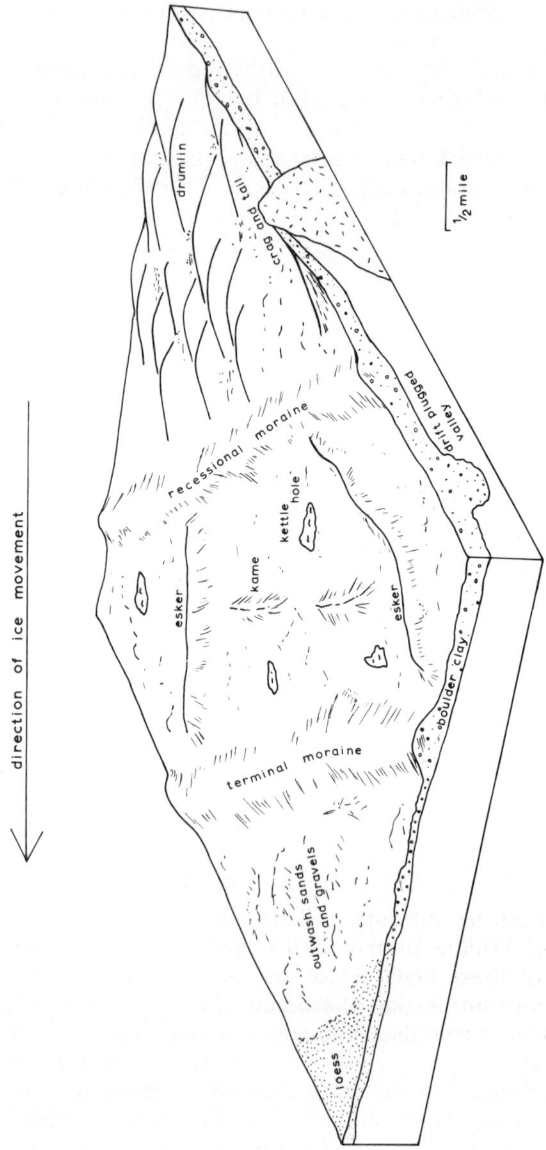

FIG. 52. Lowland glaciation

feature of groups of drumlins in most areas is the uniform orientation of their long axes. This depositional pattern is caused by the moulding action of the moving but nearly stagnant ice on its bottom or ground moraine. The long axes roughly parallel the direction of ice flow and because of that give useful information about the mass movement of the ice-sheet.

The mass wasting of stagnant ice-sheets produces many other features. For example residual patches of ice may leave small hollows or **kettle holes** which may later become small lakes. The meltwaters issuing profusely from the ice especially in the summer season lay down a great variety of fluvioglacial material. At the edge of the ice the emergent streams may deposit material in a minor deltaic fashion. If the stream is short-lived the pebbly sandy structure produced is a **kame** (Fig. 52). Sometimes the regression of the ice front is expressed by an elongated mound running at right-angles to the front and composed of fluvioglacial debris. This is an **esker** (Fig. 52) created by a persistent meltwater stream laying down material during the ice retreat. Individual eskers can be traced for many miles across country, especially in Finland.

In the British Isles the retreat phase of the Pleistocene ice was marked by the formation of extensive meltwater lakes occupying ice-dammed depressions. One of the largest was Lake Humber which covered most of the Vale of York and, to the east, the Vale of Pickering was similarly inundated. Large parts of Shropshire and the Cheshire Plain were also occupied at this time by Lake Lapworth (named after a famous geologist). Within the lakes varves (p. 39) formed. The lakes were often drained by **glacial spillways,** steep-sided valleys whose direction was controlled by the position of the ice and the topography. An example is Newtondale in Yorkshire.

Limestone Scenery

The major limestone outcrops in Britain are of Devonian, Lower Carboniferous, Middle Jurassic and Upper Cretaceous age. The areal extent of these beds at outcrop is fundamentally related to their structure but is also a function of their thickness. In the Torquay area the Devonian limestones are cumulatively thick as are the Carboniferous limestones of the Mendips, south and north Wales, the Peak District and northern England and the Jurassic oolites of the Cotswolds. The Cretaceous Chalk is a massively bedded, thick limestone throughout most of its outcrop extending from the south coast to the Yorkshire coast at Flamborough Head. In all these areas the most significant aspect of the scenery is the general lack of minor surface streams. Powerful rivers traversing the limestone outcrops and usually occupying deep gorges usually originate from outside sources. The Dove and Derwent rivers which flow through spectacular alluvium-bottomed gorges in the Peak District arise in the adjacent Millstone Grits. In passing through, very little surface water is collected from the surrounding higher land. This state of affairs is peculiar because a whole series of valleys exists on the limestones. They are dry, however, as are those on the dip slope of the South Downs Chalk near Brighton.

As nearly all the main limestone outcrops in Britain show a dry valley pattern similar to those just mentioned it means that the **water-table** (the level of permanent saturation) must have been higher in the past than at present. It would certainly have been nearer to the surface in late Pleistocene times when precipitation was higher than at present. Many of the dry valleys were probably occupied and being enlarged by streams at that time. Since then the water-table has fallen and the surface drainage has practically disappeared. In upland limestone areas the fall of water-table is greatly assisted by the permeability and high solubility of the rocks compared with sandstones and shales. The dominant component in most limestones is calcite and it is this mineral which is progressively dissolved by slightly acid ground-waters rich in carbon dioxide. The chemical reaction is expressed by the equation:

$$\underset{\text{calcite}}{CaCO_3} + CO_2 + H_2O \longrightarrow \underset{\substack{\text{calcium}\\ \text{bicarbonate}\\ \text{(carried in}\\ \text{solution)}}}{Ca(HCO_3)_2}$$

As the chemical attack tends to concentrate along joints and bedding planes they are enlarged slowly into vertical **chimneys, sinks, pot-holes** and **caverns** (Fig. 53). Hence the surface streams are gradually diverted into what may become an extensive underground network of interconnected caverns and enlarged passages, particularly with hard crystalline limestones, less so with soft chalks. In either case the water-table falls well below surface-level. At this stage original valleys are only partly occupied by streams and even these disappear in time leaving completely dry valleys. These are sometimes gorge-like as at Cheddar and The Winnats in Derbyshire. Many geologists believe that the gorges represent a long string of near-surface caverns the roofs of which have subsequently collapsed. Others suggest they are the old courses of surface streams incised along joint planes and exaggerated by late Pleistocene falls in base level.

The origin of underground caverns is also a subject of much debate. In those caves now occurring above the permanent water-table the dominant process is reprecipitation of calcite in the form of sheets, pillars, stalactites and stalagmites (p. 50). The only effect disturbing the cavity infilling are rockfalls from roof and walls. Because of this it has been suggested that cave formation must occur either below or at the level of the water-table, providing the latter remains stable at one level for several thousand years. Enlargement is probably assisted by the abrasive

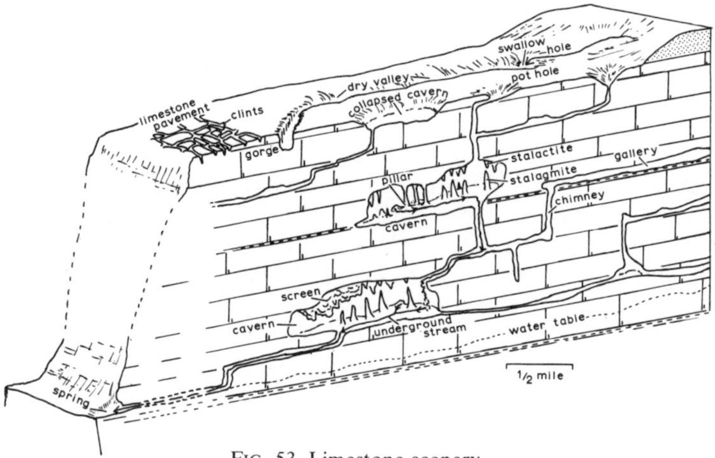

FIG. 53. Limestone scenery

action of clastic loads within the circulating waters. Fluvial action must be usual because layers of alluvium and river gravels often occur in the caves. Sometimes they are mantled by younger thick layers of multi-coloured spar (calcite) in which traces of prehistoric and historic animal life are found. The Cresswell Caves in the Magnesian Limestone (Permian) of east Derbyshire have yielded evidence of Pleistocene elephants, rhinoceros, hippopotamus and Roman man and Kent's Cavern at Torquay in Devonian limestones the bones of Pleistocene woolly rhinoceros, reindeer, mammoth, sabre-toothed tiger and Palaeolithic man.

Apart from features such as sink-holes and dry gorges the most typical characteristic of hard limestone country is the general cragginess of the topography with abundant bare scars outcropping from thin soils. Exposed bedding planes, moreover, are fretted by solution along joints to form deep grooves separated by narrow ridges of rock. These fluted pavements are called **clints** or **grykes** (Fig. 53). The broad scenic effect is well developed in the Derbyshire Dome but is even more apparent in the Causses Plateau region of south France or the Dinaric Alps of Yugoslavia. In the last two areas the scenery is so stark, the soil so generally poor and the underground drainage so extensive that the name **karst** is used to describe them. In some parts the solution effects seem to have been so extensive that several square miles of limestone surface has collapsed forming flattish depressions called **poljes.** In Derbyshire the scenery shows only the small-scale features of a fully-developed karstic landscape; large features such as poljes are absent. The same can be said about the Mendips and parts of the northern Pennines. The Chalk areas of Britain are not karstic either despite the presence of dry valleys, sink-holes and thin soils. The Chalk is too soft to form crags and extensive interconnecting underground caverns are unknown.

Desert Scenery

At the present day deserts cover approximately 20 per cent of the world's land surface. In all of them the rainfall is low (less than 6 in. per year) and evaporation high, consequently there is no continuous cover of natural vegetation. As clouds are rare the diurnal range in temperature may be considerable, often more than 100°F. (38°C.) during the day and less than 32°F. (0°C.) at night. This is particularly so in the hot subtropical deserts such as the Sahara, Arabian and central Australian, but is almost equally true for those of high latitudes occurring in the 'rain shadow' of mountain ranges. Some of the latter are found within the Rocky Mountain system of North America, others within and immediately to the north of the Himalayan chain. All are arid as the clouds cannot rise above the protecting mountains without releasing most of their rain.

Adjacent to all true deserts there are semi-arid steppe areas which have transitional characteristics and grade into regions of persistent vegetation and regular though often seasonal rainfall. Somewhat oddly the steppes commonly show evidence of extreme aridity in the past and it can only be concluded that the world's climatic belts have altered their position slightly during the last million or so years. The movement was probably linked with the presence of extensive circumpolar ice-sheets during Pleistocene times, altering the position of the earth's atmospheric pressure belts. Conversely, many of the present true deserts have erosional valley forms which seem to have been partly created by higher rainfall than now. This also probably results from Pleistocene climatic changes.

Despite the low average rainfall and the greater depth of the present water-table than in wetter regions, the work of water is very important in the development of desert land-forms. This is because the rainfall is usually localized and very heavy. More than 6 in. may fall in one day followed by a year or so of complete drought. Consequently, torrential streams may gush down gullies, valleys and **wadis** (flat-bottomed, cliff-sided valleys) for only a few hours. The rates of erosion, run-off and transportation are speeded up by the lack of vegetation cover so that during the short downpour the erosion accomplished is much

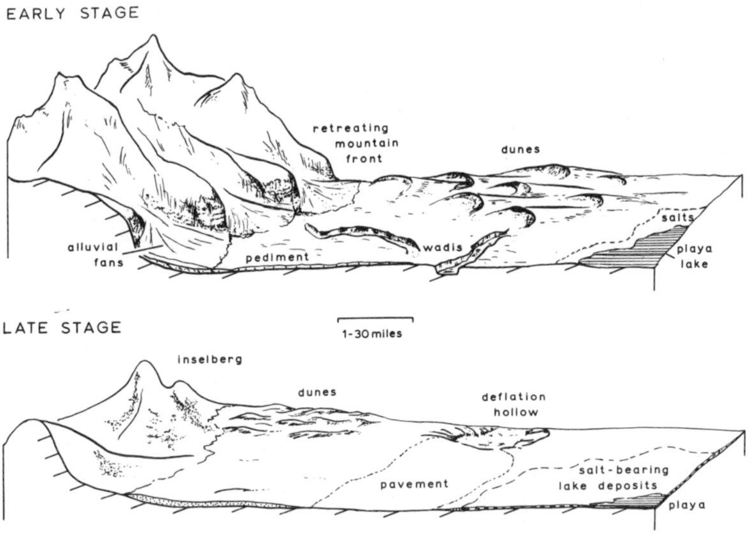

FIG. 54. Desert scenery

more than that of a stream in humid regions which has been flowing for many years. The torrents (or flash-floods), especially those issuing from more hilly areas, are abnormally laden with sand and silt. On reaching flatter lower levels the velocity of these sediment-choked streams falls quickly and the load is laid down as alluvial fans at the margins of the high land. Simultaneously, the solution load and finest suspended matter may be deposited in small temporary lakes in the lowland. These are called **playas** (Fig. 54) and within them salts such as halite and sodium carbonate are precipitated. Extensive alluvial and minor evaporite deposition, therefore, typically characterizes true deserts. The feeder streams never persist sufficiently long for a unified drainage pattern to develop as in more humid areas.

The effects of rainfall are intermittent but there are slower processes which proceed without break. Thermal expansion and contraction and capillary ground-water movements combine to physically disintegrate rocks. Exfoliation is commonplace especially along bare, rocky, fault-bounded mountain range fronts, where the loosened debris progressively accumulates to create screes. The screes merge into adjacent alluvial fans and in due course a clastic apron develops running along the full length of the mountain front. At spasmodic intervals the debris is swept away by flash-floods and fresh rock faces exposed to physical decay. The effect of this on the mountain front is to make it retreat and create simultaneously at its foot a nearly horizontal surface known as a **pediment** (Fig. 54). Some of the older pediments of deserts are many miles wide and their solid rock surfaces are usually mantled by a thin layer of redistributed scree and alluvial debris. The final stage of mountain reduction is seen in old deserts such as the Kalahari of Africa where isolated **inselbergs** (Fig. 54) rise out of the pediments like islands in a rock sea.

Wind action is important in all deserts but is not necessarily the dominant process at work. Water is much more active as an eroding agent and more is accomplished by a short torrential downpour than by many years of wind erosion. The main effect of wind is to pick up the loosened sand debris of fans and pediments and transport it elsewhere, often a few thousand yards, more rarely many hundreds of miles. Loess (p. 102) is a good example of the latter. During duststorms the wind with its whirling load of suspended sand becomes mildly abrasive in nature and bare rock outcrops or man-made features are attacked. Panes of glass gradually become frosted as do most of the individual grains of sand which are actually doing the frosting. In fact one of the ways of distinguishing wind-blown sand from water-deposited is to look for the dull frosted surface on the quartz grains. If this is done it can also be seen that the grains are dominantly well-rounded owing to mutual attrition during transport. Apart from smoothing rock surfaces the rasping wind also picks out the less resistant layers, especially those at heights less than 6 ft. from the ground. Very little sand is carried above that height by normal desert winds. **Zeugen** (Fig. 55) are pedestals of soft rock 5 to 15 ft. high with a hard rock capping. **Yardangs** (Fig. 55) are near-parallel grooves up to 20 ft. deep separated by upstanding hard ribs and extend lengthwise for many hundreds of yards. In their etching the prevailing wind direction, dip of the strata and trend of major joints probably play an important role.

The intermittent blowing of silt and sand away from any locality means that the surface may be progressively lowered either a few inches or, sometimes, many tens of feet. The process is called **deflation** and it has been suggested that many large depressions in deserts are created by it. The Qattara depression on the borders of Egypt and Libya occupies an area of approximately 10,000 sq. miles, most of which is below sea-level. Its northern edge is a steep but strongly dissected escarpment whereas to the south it rises gently to merge with the surrounding desert. On the southern side are extensive **dunes** of sand apparently excavated from the depression to the north. In all depressions of this kind wind erosion ceases at a certain level related to the presence of the underlying permanent water-table. Below this level the sand grains are firmly welded together by pore water. A thin cover of vegetation is often present too. The more densely vegetated parts surrounding freshwater springs are the well-known oases. In depressions and their surrounding more arid areas the action of the wind tends to separate the sand from the pebbles and larger boulders. The sand is wafted away and the larger fragments remain behind as a residual surface deposit known as a **desert pavement.** An examination of the residual fragments is interesting as many of them show flat, straight-edged faces (or facets). This flattening is caused by sand abrasion.

Probably the best known but not necessarily dominant feature of deserts are sand dunes made up of grains greater than 0·06 mm. in size; below that size the grains are too small to build up into mounds. The type of dune scenery eventually formed depends on at least three interdependent factors: vegetation, sand supply and wind speed and direction. Vegetation acts as a physical barrier to sand movement and as a stabilizer by active growth on the dunes. In the rockier parts of deserts the sand supply may be low and may easily be obstructed in movement by scrub. The sand tends to accumulate against the scrub patches and slowly a group of separate dunes forms, each rising to a height of several feet. At this stage under uniform wind conditions the dunes develop an asymmetric cross-section (Fig. 55) with shallow slopes on the windward faces and steep slopes on the leeward (or slip) faces. Simultaneously the wind deflection around the dune sides drifts loose sand so that each has two forward-facing arms. The crescentic structures are now known as **barchan dunes.** As they increase in size the vegetation tends to be inundated and if the winds remain steady they begin to migrate downwind. Whole groups of barchans may move for many tens of miles covering and effectively destroying any objects in their path. The same

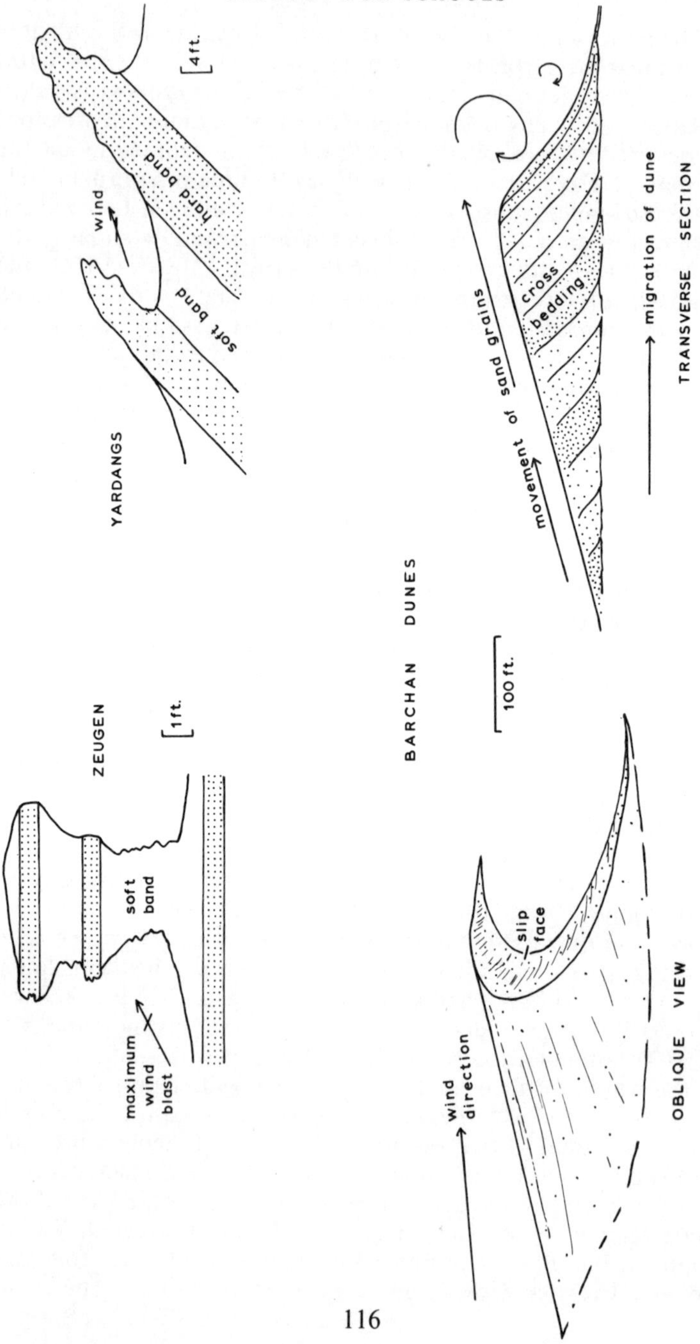

FIG. 55. Yardangs, zeugen and barchan dunes

is true but on a grander scale when the original supply of sand is high. Then **sand seas** develop covering hundreds of square miles and composed in part of migratory **transverse dunes** running at right-angles to the wind direction for distances of up to a mile. With high wind velocities and low sand supply **longitudinal dunes** paralleling the wind direction may form. Groups of them may persist for many tens of miles with individual heights anything up to 30 ft. On the surfaces of all types of dunes **wind ripples** are generated during periods of low wind velocity. They are invariably asymmetrical and have wave-lengths of almost 7 cm. and amplitudes of about 3 mm.

The detection of ancient desert environments is much more difficult than might appear from the knowledge we have of present-day processes and effects. In Britain the main period of time during which the evidence favours desert conditions is the Permo-Triassic (p. 163). Salt deposits of many kinds were precipitated, especially in Cheshire and northern England. Some were almost certainly laid down in playa lakes. Many of the sandstones appear to be fossilized dunes although true wind ripples are very rare. The sand grains are well rounded and frosted and cementation by haematite is common which suggests capillary action in hot, arid or semi-arid climates. There are pebble beds too which resemble desert pavements and, as would be expected, there are very few traces of vegetation.

QUESTIONS

1 Distinguish between weathering and erosion. Describe the effects of erosion within and in the vicinity of rivers after a eustatic fall in sea-level.
2 Account for trellis drainage patterns.
3 Describe and account for features associated with the retreat of lowland ice-sheets.
4 Why are water-tables usually low in areas of thick limestone outcrop? What features are there which indicate a higher water-table in the past?
5 Describe the rocks and features you would expect to find preserved in fossil desert landscapes.
6 Write a short account of modern ideas on off-shore shelf sedimentation.
7 What indication is there around present-day coasts that waves have the capacity to erode?

Chapter 6

FOSSILS

THE scientific study of ancient life as shown by the preserved remains of animals and plants is known as **palaeontology.** The actual organic remains or traces are **fossils** and it is to these that the geologist turns when reconstructing the life patterns of land and sea during former ages. The work is not easy because the fossils have to be extracted carefully from the sedimentary rocks and then identified correctly. There are hundreds of thousands of named varieties of fossil and great skill is needed for precise identification of any one individual. Complications are introduced into the distribution patterns of ancient life by the fact that many skeletons are destroyed either by scavenging organisms or by normal erosion processes. On average probably not more than one skeleton in 10,000 has been preserved in old sediments and nearly all those have lost all trace of soft parts. A biologist uses soft and hard parts in classifying present-day life, but the palaeontologist has not this advantage and has to group his fossils mainly on the basis of hard parts. It is hardly surprising that some extinct fossil groups are difficult to classify and relate to modern creatures. The graptolites are a prime example of this (p. 141).

The processes of fossilization are varied and sometimes complicated. Only rarely are remains preserved unaltered and invariably this is due to freakish conditions. Whole Tertiary flies, for example, have been entombed in the resin of trees, Pleistocene woolly mammoths have been deep-frozen in permafrost areas of Asia and a wide range of Pleistocene to Recent animals in an asphalt lake in Trinidad. Some chitinous shelled brachiopods ranging in age from Cambrian to Recent times also seem to have altered little with the passage of time. Probably the commonest process is alteration by replacement. This is effected by circulating ground-waters rich in minerals and can take the form of calcite skeletons being replaced molecule by molecule by chalcedonic silica. In limestones where much secondary silicification has occurred the altered organism may be preserved in perfect detail. Pyritization has the same effect as do replacements by dolomite, siderite and haematite. Moving ground-water also has the capacity to dissolve away completely skeletons embedded in rocks and by doing so a cavity or **mould** is created, on the sides of which are imprinted the surface ornamentation of the original fossil. If the mould is then infilled with secondary minerals a **cast** is formed which very closely resembles the original in all respects except chemical composition. Replaced fossils, moulds and casts can be extracted from rocks and accurately named but there are many organic structures preserved where the organism itself has completely disappeared. Many of these **trace fossils** are tube-like structures of varying attitudes and diameters possibly caused by burrowing worms. Other traces suggest trilobite and gastropod trails and some are clearly the footprints of animals such as dinosaurs.

The extraction and subsequent treatment of all types of fossil is of prime importance. Their value becomes almost negligible, however, if they are not accurately localized geographically and geologically. It is essential that the site of discovery and specific bed containing the fossil should be noted. When this information is known the process of naming can begin. First it is decided whether the organism belongs to the animal or plant **kingdoms,** and if the former, then to which particular major group or **phylum.** There are 14 animal phyla distinguished by geologists, some of which need not be considered here. The important ones are:

	PHYLUM
Animal Kingdom	Coelenterates (Corals) Echinoderms (Crinoids, Echinoids) Brachiopods Molluscs (Lamellibranchs, Gastropods, Ammonites, Belemnites) Arthropods (Trilobites) Hemi-Chordates (Graptolites) Chordates (Vertebrates; Fish, Amphibians, Reptiles, Birds, Mammals)

Having placed the fossil into its correct phylum the final stage is to give it a **generic** and **specific** name. All fossils have double names. The first name or **genus** always begins with a capital letter whereas the second name or **species** has no capitals. For example, a well-known echinoderm from the Upper Chalk of south-east England is called *Micraster* (genus) *cor-anguinum* (species).

All the phyla have a considerable time range (Fig. 56) and most extend through from earliest Palaeozoic times (600 million years ago) to the present day. If individual groups of organisms within each phylum are examined a more accurate picture of events is obtained. Some begin in a tentative way, proceeding to a big burst of activity during which many genera and species are distributed over wide areas and, finally, completely die out for no obvious reason. The ammonite group (Cephalopods) after reaching a peak in varieties during Mesozoic times are now extinct. So are the graptolites (Hemi-Chordates) of Lower Palaeozoic times and so are many groups of vertebrates. The discovery of coelacanth fish in the Indian Ocean in the 1950's created a stir in the scientific world as the group was thought to be virtually extinct. Then there are groups or varieties of organism such as *Lingula* (Brachiopod) which persist for great lengths of time without any significant change. The Lingulas of Ordovician times differ little from those of present times.

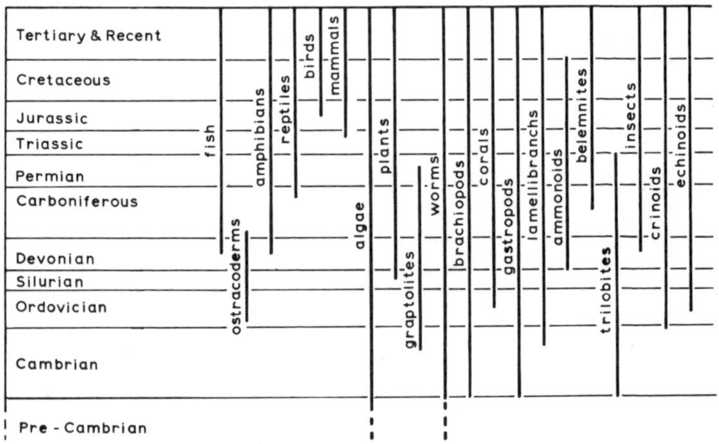

FIG. 56. Range of fossil groups

Coelenterates

Most members of this phylum whether recent or fossil are marine in nature and many secrete calcareous skeletons. The best-known modern representatives are sea anemones, corals and jelly-fish, whereas corals are the most important fossil group.

Corals

The conditions under which active coral growth takes place are well known. Shallow clear waters (under 300 ft. deep) and warm marine currents (above 68°F.) are essential for coral polyp growth in reefs, though non-reef corals can exist at greater, muddier depths (up to 600 ft.) and much lower temperatures. These conditions are assumed to have applied in the past. The earliest corals known are Ordovician, but it was not until Silurian times that they developed in sufficient numbers to become widespread throughout the world. The first large coral-reefs are also Silurian and are particularly well developed in Shropshire, south Sweden and the Great Lakes area of North America. From that time onwards corals have always grown in relative abundance on the sea-floor, providing the environmental conditions have been suitable.

The habit, shape and size of corals are partly dependent on environmental factors. If the conditions are a little extreme and unsuitable for reef growth then a **simple** (solitary) habit is common; the organism exists as an individual isolated from adjacent forms. Under reef conditions the habit is normally **compound** (colonial) with the individual **corallites** closely intergrowing and even welding one to the other. Fig. 57 illustrates two common genera with these contrasting habits. *Dibunophyllum* (Fig. 57) is a simple coral of Carboniferous age and is usually found in limestones. An average size for British examples is 1 in. in maximum diameter and 6 in. in length. The end embedded in the bottom limey muds is conical but the shape quickly becomes cylindrical upwards as the coral polyp secretes more calcareous skeletal matter. The outer skin (**thecal wall**) is rough and ridged and because of this the coral is a member of the **Rugose** group. When horizontal and vertical sections are taken through a specimen a number of important internal structures are revealed, all of which contributed to the rigidity of the coral skeleton. The most prominent are numerous radial **septa** which are thin vertical plates commencing at the outer wall and running inwards for variable lengths towards the **central column** or **columella.** Some columellas are flat plates running axially through the full length of the coral as with *Dibunophyllum,* others are more like a solid rod, e.g. *Lithostrotion* (Carboniferous). Adjacent to the theca and supporting the septa are numerous small curved plates called **dissepiments.** The width of the outer zone (**dissepimentarium**) occupied by these small plates in *Dibunophyllum* is about one fifth of the radius but in other corals it may be much wider. *Palaeosmilia murchisoni* (Carboniferous) has a dissepimentarium extending nearly two thirds of the way to the centre which, incidentally, has no columella. Across all these named varieties of coral from theca to columella run a series of near-horizontal plates called **tabulae.** They are thin calcareous layers secreted by the polyp during pauses in its upward growth.

Whereas *Dibunophyllum* has a large variety of internal plates, other corals have little. *Zaphrentites* (Carboniferous) has no columella or dissepiments. *Montlivaltia* (Trias to Tertiary), a **Hexacoral** with septa arranged in multiples of six, has no tabulae or columella.

Colonial corals were much more widespread in Mesozoic times than during the preceding Palaeozoic era. By this time too the

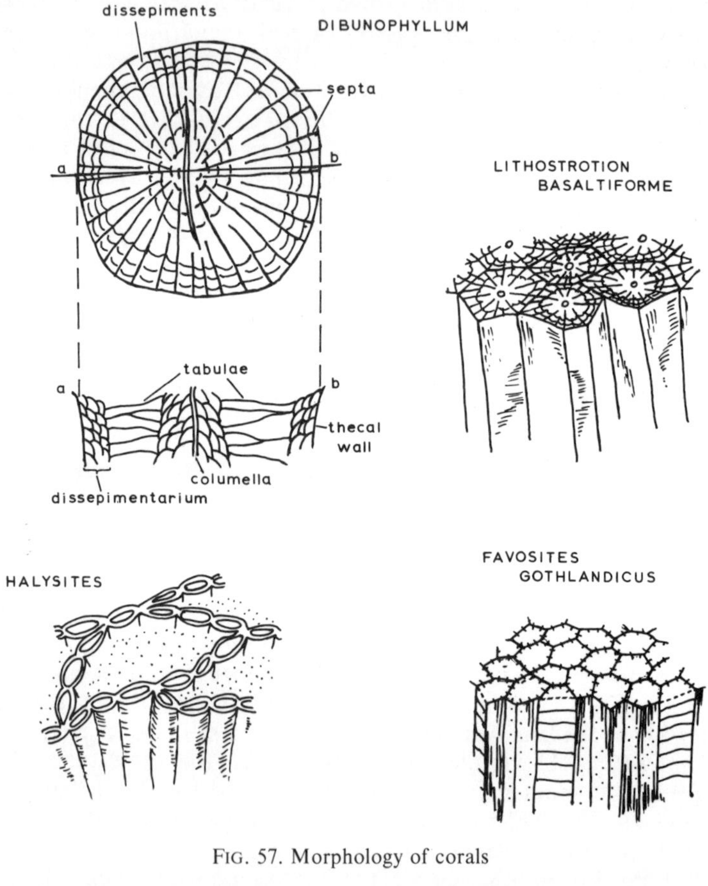

FIG. 57. Morphology of corals

Rugose corals were extinct and had been replaced by the Hexacorals. None the less some of the best-known colonial genera occur in the older rocks. *Lithostrotion martini* (Carboniferous) is one in which the individual tubular corallites are joined and intergrown in the manner of branches on a tree. The habit is created by a budding or splitting (fission) process during growth. Occasionally the multiple growth is such that the corallites weld together along their length and in doing so each develops a polygonal cross-section. *Lithostrotion basaltiforme* (Carboniferous) (Fig. 57) is of this kind as are *Isastraea* (Triassic to Cretaceous) and *Thamnasteria* (Jurassic). The habit is also characteristic of *Favosites gothlandicus* (Silurian) (Fig. 57), a member of the now extinct **Tabulate** group of corals. Another Tabulate *Halysites* (Silurian) has a chain-like colonial habit (Fig. 57). The Tabulate corals have tube-like skeletons with closely-spaced tabulae; they differ from the rugose corals in not having wrinkled thecal walls and not having well-developed septa.

Importance of Corals

The Rugose corals evolved rapidly during Palaeozoic times and were widespread in clear and muddy marine environments. The Hexacorals appear to have evolved from them in early Mesozoic times and since then have increased to such an extent that they are the dominant form in modern seas and oceans. The origin of the Tabulates is doubtful but they were at their peak of development in Palaeozoic seas, were less widespread in the Mesozoic and eventually became extinct. All groups are marine and some varieties were capable of building reefs, so the original environment can be deduced by studying them and their containing rock. However, because corals are specialized in their environmental needs and are fixed to the sea-bottom during life their value in dating and correlating beds is poor. In Carboniferous times, for example, the same coral genus *Zaphrentis* occurs in beds of different age. The beds are similar in lithology, despite the age difference, so it is certain that the presence of the coral is related more to environmental factors than to the age of the rocks. Some of the colonial species occurring in fossil and modern reefs become less abundant when traced away from the reefs and eventually may become subordinate to different species of simple habit. In old strata this means that correlating reef structures with adjacent rocks deposited simultaneously is very difficult if only the corals are examined.

Echinoderms

The echinoderms are marine organisms with calcareous skeletons and have a wide time range extending from the Cambrian up to the present day. The commonest feature of all the varieties is a bilateral and pentameral symmetry which is expressed in the skeleton either by patterns of plates or by the presence of five arms. Basically the echinoderms are scavenging creatures living on the dead and decaying bodies of anything raining down on to the sea-bottom. Some also filter the top layers of sand and mud through their stomach, clearing them of all organic debris. The sea lilies or **crinoids** are, generally, a passive group as many of them spend part of their life attached to the bottom. The sea-urchins or **echinoids** are capable of locomotion along the sea-floor and can seek out their food. So can the starfish or **stelleroids.** The last two are usually protected during life by coarse tubercular or spiny surface outgrowths.

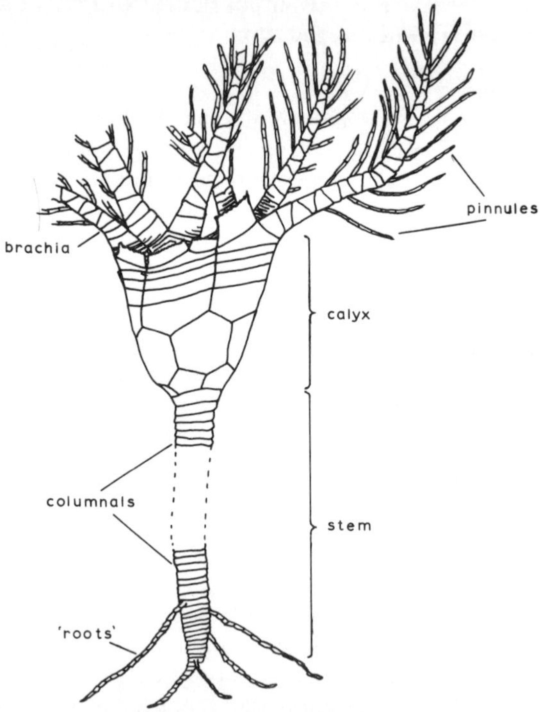

FIG. 58. Morphology of crinoids

Crinoids

Crinoids, especially the stalked species (Fig. 58), live in colonies on shallow sea-floors. They occur profusely in the vicinity of reefs and hence, in the geological column, are repeatedly found within marine limestones. Sometimes the broken-up fragments of dead individuals are swept into low mounds or crinoid banks by localized currents. The earliest known crinoids are Ordovician in age. They reached their peak in Silurian to Carboniferous times and since then have slowly declined. Most of the older forms, e.g. *Cupressocrinus* (Devonian), had stems and adhered to muddy or rocky bottoms whereas from Jurassic times onwards the proportion of unstemmed free-floating forms, e.g. *Marsupites* (Cretaceous), has increased until they form over 80 per cent of all living species.

The skeletal morphology of a generalized crinoid is illustrated in Fig. 58. The stem or column, when present, consists of a number of articulated discs called **columnals** which allow the organism to bend and withstand buffeting by currents. The discs vary in cross-section, some being circular, others four- and five-sided. At the base of the stem there are 'roots' and at the top a cup (**calyx**) formed of larger calcareous plates which protected the living crinoid. The roof of the calyx has a covering of plates in which are two holes, one being the mouth and the other the anus. At the margins of the calyx are attached five arms (**brachia**) each consisting of tapering rows of plates. Down each arm run food-grooves leading to the mouth. In many cases the food is guided into the grooves by rows of small supplementary arms known as **pinnules.**

Importance of Crinoids

Crinoids are valuable indicators of specialized marine environments but the fixed species, being rather limited in areal distribution at any given time, are of less use for correlation purposes. Free-floating species can be used in correlation, e.g. *Marsupites testudinarius* in the Upper Chalk of Britain.

Echinoids

Echinoids have existed from Ordovician times to the present day. They became numerous in Carboniferous times and from then on have always formed an important part of the total fauna in marine sediments, especially limestones. In contrast with crinoids they have no arms or stems and their mouths are downward-facing, suitable for the bottom-grubbing mode of life. The skeleton of echinoids can be roughly described as a hemisphere with the flat **oral** side downwards (Fig. 59). This shape is well developed in the **Regular** group but less so in the **Irregular** group since there is distortion related to differing positions of the mouth and anus.

Irrespective of shape all echinoid skeletons have a basic similarity of structure. They consist of three parts: **corona, peristome** and **apical disc.** Forming most of the shell is the corona which consists of a radial pattern of two types of rigid interlocking plates. There are five zones of **ambulacrals (ambs** for short) which alternate with five zones of **interambulacrals** (**interambs**). The ambs consist of small plates with pores out of which **tube feet** protruded during life. These 'feet' help the animal to move and hold itself firmly on the bottom and probably help to channel food towards the mouth. The interamb plates are larger and often marked by **spine bases** to which **spines** were attached during life. The spines may be large club-shaped objects as in *Cidaris*

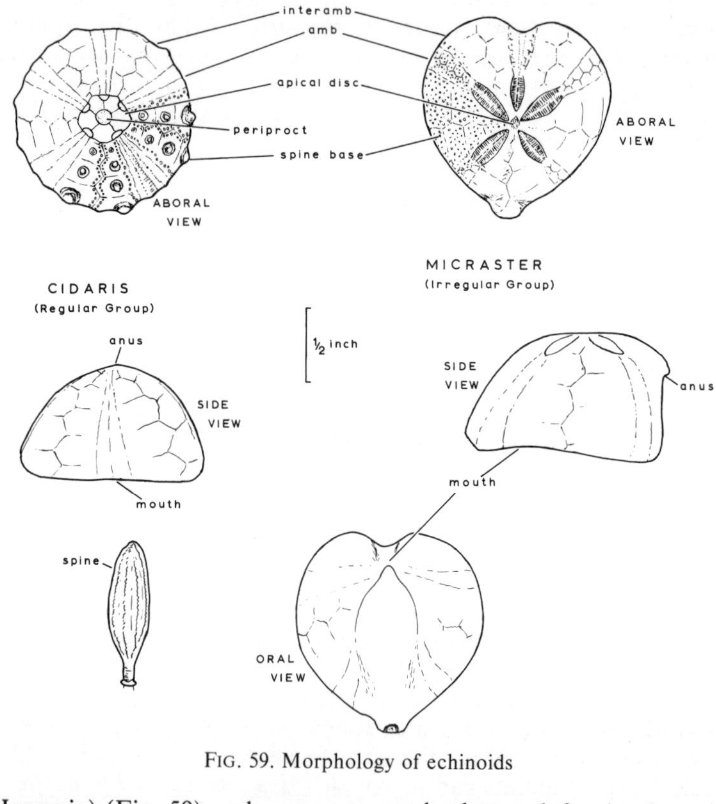

FIG. 59. Morphology of echinoids

(Jurassic) (Fig. 59) and appear to act both as a defensive feature and a means of locomotion. In contrast *Micraster* (Cretaceous) (Fig. 59) has fine hair-like spines. At the focus of the ambs in the Regular echinoids, e.g. *Cidaris*, there is the mouth on the bottom **oral** side and the anus on the top **aboral** side. In life the mouth of *Cidaris* is surrounded by many small plates and the whole structure is then called the peristome. In other echinoids the peristome has five pairs of plates around the mouth. The anus is similarly complex in life and forms part of the **periproct.** In

Regular echinoids the apical disc consists of a ring of ten plates surrounding the periproct (a membrane covered with small plates) which in turn surrounds the anus. During the evolution of Irregular echinoids this simple pattern of plates has changed. The periproct, anus and one plate move out of the apical disc and down the posterior interamb. This is well seen in *Micraster* (Fig. 59); its mouth has also moved forward and is protected by a lip. *Holectypus* (Jurassic) shows the extreme development of the trend of the movement of the periproct since it appears on the oral surface. This echinoid is flattened with a circular outline, well adapted for its life in shallow water.

Importance of Echinoids

As echinoids tend to exist in colonies their value in correlating adjacent beds whether of similar lithology or not is poor. In certain instances, however, they are used because there are no good alternatives. The Micrasters and Holasters are valuable in this sense in correlating the Chalk outcrops of Britain.

Brachiopods

Brachiopods are bi-valved marine organisms inhabiting a wide variety of sea-bottom environments generally at depths of less than 600 ft. Some are adapted for muddy in-shore and deep-water habitats and tend to be thin-shelled. Others have a ribbed, thick shell and can withstand battering by powerful waves and currents, e.g. *Rhynchonella* (Jurassic to Cretaceous) (Fig. 60). In some clear limey seas in the past the conditions have been so perfect that the rock produced consists of very little else but brachiopod shells. Parts of the Aymestry Limestone (Silurian) of Shropshire are formed of thick shell-banks of *Conchidium knighti*, a shallow-water, coarse-ribbed variety.

Brachiopods reached peak development in Lower Palaeozoic times and have progressively declined in varieties and numbers up to modern times. The earliest known are Cambrian, e.g. *Lingulella*, so it is almost certain that the first members of the phylum originated in the late Pre-Cambrian. Some of the very old genera have rather primitive characteristics, particularly a lack of interlocking teeth at the posterior end of the two valves. They are named the **Inarticulata** and some varieties have managed to persist through to the present day. However, most fossil and modern genera have teeth and belong to the **Articulata** group.

The shells of brachiopods are calcareous, phosphatic or chitinous and consist of two unequal-sized valves, **ventral** and **dorsal** (Fig. 60). The valves completely surround and protect the living organism which feeds by opening its valves along the anterior edge (**commissure**). At this end also waste products are ejected and in some genera the shape of the valves is modified by an upward **fold** or downward **sulcus** (**sinus**) to ease these processes. Outer surfaces of valves vary in ornament, some being smooth, others having **ribs** radiating outwards from the posterior end. Additionally the valves may be corrugated by concentric **growth lines.** If a median plane is drawn at right-angles through the hinge and anterior margins of either valve it can be seen that each half is the mirror-image of the other. The valves are bilaterally symmetrical. The ventral valve is usually the larger of the

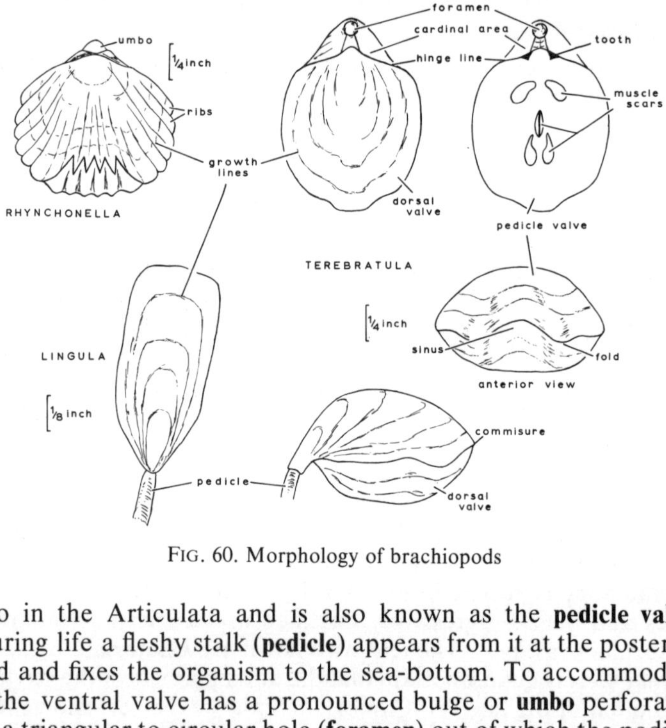

FIG. 60. Morphology of brachiopods

two in the Articulata and is also known as the **pedicle valve.** During life a fleshy stalk (**pedicle**) appears from it at the posterior end and fixes the organism to the sea-bottom. To accommodate it the ventral valve has a pronounced bulge or **umbo** perforated by a triangular to circular hole (**foramen**) out of which the pedicle extends. A few genera in the past seem to have lost the use of their pedicle and foramen. *Productus giganteus* (Carboniferous), which has a width of up to 1 ft., is of this type and fixed itself to the bottom by long spines growing out of the ventral valve.

Abutting the edge of the foramen in brachiopods and sometimes running externally along the full length of the **hinge-line** is a flattish zone called the **cardinal area.** In *Spirifer* (Devonian to

Permian) the area is relatively wide whereas in *Terebratula* (Devonian to Recent) (Fig. 60) it is often difficult to see at all. The smaller **dorsal valve** also has a cardinal area. This valve differs in many ways from the ventral. There is no pronounced umbo and no foramen, and teeth sockets occur at the hinge-line instead of teeth. In the position analogous to that of the foramen in the ventral valve there is often a ridge or protuberance called the **cardinal process.** This functions as an anchorage for some of the muscles of the living creature and gives the necessary leverage for opening and closing the valves. The relative size of these muscles is sometimes indicated on the inside surface of both valves by a number of depressions called **muscle scars.** In addition to scars the insides show many kinds of umbonal structures which strengthen and support the creature during life. Although these structures are important in classification the terminology is too complicated for further description.

Importance of Brachiopods

Brachiopods are interesting to the geologist as they are so widespread in differing kinds of marine environments. *Lingula* (Ordovician to Recent) (Fig. 60) prefers muddy shallow water conditions, others such as *Productus* (Carboniferous to Permian) could exist in either muds or in clear limey seas. As some genera were tolerant of a variety of conditions they can be used for correlating dissimilar rocks and this compensates for the colonial and attached habit of most varieties.

Molluscs

The phylum Mollusca is one of the largest and most important in the geological column. The number of species has progressively increased since early Cambrian times and more than 60,000 living species are known. The ability of the group to adapt itself to differing conditions has been so great that there are very few land or sea areas which have no representatives at present. The best-known oceanic varieties are giant squids and the octopus, in shallower shelf seas and estuaries oysters, mussels and clams, and on land snails. The variation in size and shape of the living creature and its skeleton are equally as varied as the environments they inhabit. Squids may be 50 ft. long, snail shells 1 mm. Then some molluscs are single-valved, others double, some thin-shelled, others thick. On the basis of shell character the phylum can be subdivided fairly easily into many distinctive groups. The most important of these are the lamellibranchs, gastropods, ammonites and belemnites.

Lamellibranchs

The bivalved lamellibranchs occur in a wide range of fresh, brackish and salt-water environments. The shells, which are usually calcareous, are modified according to the habitat. If there is an abundance of mud and slow sedimentation, as there might be in an estuary, then they are commonly thin-shelled. In more agitated marine waters they tend to be thick-shelled or strongly ribbed. The common oyster (*Ostrea*) of East Anglian estuaries has a thin but strong laminated shell which appears to adhere and mould itself to the muddy surface. Because of this particular mode of life the two valves of oysters are markedly different in size. This is unusual in the lamellibranchs but can also be observed with *Gryphaea*, *Exogyra* and *Inoceramus*, all of which were common in Mesozoic times.

Ignoring specialized genera such as those just mentioned, the main property of lamellibranch shells which distinguishes them from other organisms is two equal valves (Fig. 61). A plane of symmetry can be drawn between them. The plane passes between the umbos which are usually of equal size in each valve. The valves are called **left** and **right** as there is no pedicle. Correct orientation is attained by looking down on to the umbos and hinge-line with the anterior end pointing away from the viewer. The anterior end is sometimes shorter and more rounded than the posterior although there is no general rule allowing ease of orientation. Most of the Upper Carboniferous non-marine lamellibranchs, e.g. *Carbonicola* (Fig. 61) have their umbos nearer to the anterior margin. In other genera the position of internal scars may be useful.

The hinge-line of nearly all lamellibranchs is characterized by interlocking teeth and sockets. Each valve has a set and they are useful for classification purposes. External and adjacent to the hinge are cardinal areas which are hidden in life by a thick elastic **ligament.** The ligament functions as a spring, forcing the two valves apart and allowing free circulation of water around the internal organs. To counteract this springing apart of its valves the living organism has sets of internal muscles (**adductors**) and these pull the two valves together. The position of the adductors is shown by scars in each valve. Most genera, e.g. the common cockle, *Cardium*, have two sets, one at the anterior and the other at the posterior (Fig. 61). This is the **dimyarian** state. *Ostrea* (Jurassic to Recent) has only a posterior set as does *Pecten* (Silurian to Recent). This is the **monomyarian** state. Near to and parallel to the ventral margin of both valves and running from the anterior to the posterior are **pallial lines** which mark the position of attachment of the living mantle to the shell. In some genera the line is quite smooth, e.g. *Glycimeris* (Pliocene to Recent), whereas in others a posterior indentation known as a **pallial sinus** occurs, e.g. *Mya* (Oligocene to Recent) (Fig. 61). This is against a permanent opening (gape) in the two valves.

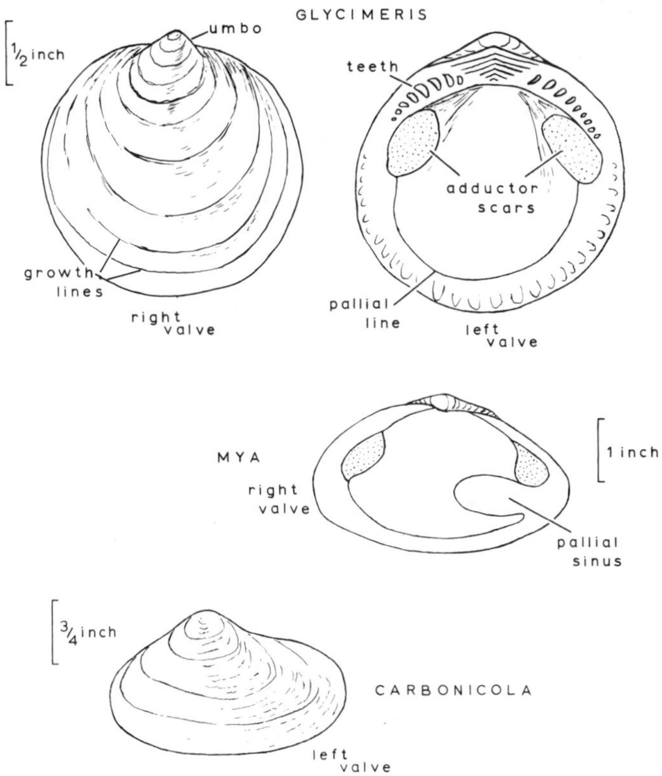

FIG. 61. Morphology of lamellibranchs

Importance of Lamellibranchs

The lamellibranchs live entirely in water, are mainly marine, and are usually attached to the bottom either by cementation or by a series of thin horny threads called the **byssus.** A few varieties, such as *Teredo* (Jurassic to Recent), the 'ship-worm', bore into wood and rock and become permanent residents, others burrow in muds, e.g. *Mya*, and yet others appear to float freely at least for part of their life, e.g. *Dunbarella* (Carboniferous). The rapidity with which they have adapted themselves to different habitats is useful, therefore, for environmental interpretation. It is less good for correlating strata of equivalent age but differing lithology. Correlation is complicated further because the two valves tend to part on death of the organism and become more susceptible to distortion by compaction processes. Until Devonian times very few lamellibranchs seem to be well preserved. From then on preservation has improved and assemblages of shells have been used in Carboniferous coalfield correlation. Oysters, e.g. *Ostrea delta* (Jurassic), are used in parts of some Mesozoic successions and *Inoceramus labiatus* is the index fossil for part of the Middle Chalk of south-east England. The Tertiary rocks have a greater abundance of lamellibranchs than any preceding era and hence are used much more freely in correlation.

Gastropods

Gastropods form the largest group of the Mollusca and have progressively become more abundant since Cambrian times. At present nearly 30,000 living species are known. The earliest Palaeozoic forms, which are poorly preserved, have a simple external ornament and seem to have been nearly entirely marine. They breathed by means of gills. By Mesozoic times the gastropods had conquered fresh waters and land-surfaces and were widespread in both. The latter adaptation was accomplished by the development of lungs in place of gills. Even on land-surfaces varieties such as snails and slugs preferred and still prefer damp calcareous soils with a good cover of vegetation. The vegetation is essential because most gastropods eat plants; marine gastropods eat seaweed. Gastropods, in general, are free-moving animals and this allows them to seek food wherever it is available. Their teeth which are rasp-like are so powerful that other shelled creatures can be attacked. The common whelk *Buccinum* (Tertiary to Recent) eats through lamellibranch and brachiopod shells to feed on the soft parts within.

Despite their great variation in shape, size, colour and ornamentation gastropods are classified by reference to their soft parts. With modern species this is a simple process but the older the species is the more difficult it becomes. Fortunately a few of the really old Palaeozoic genera closely resemble modern forms and are classified accordingly. Others such as *Hyolithes* and *Tentaculites* which occur in Lower Palaeozoic rocks of Britain have no modern equivalent and may not be gastropods at all.

Gastropod shells are univalve and usually calcareous, irrespective of environment. The valve is normally only partly occupied by the creature which can emerge from and retreat into its living-chamber when necessary. The valve may be a simple inverted round-topped cone as in the common limpet *Patella*, a flat spiral (plano-spiral) or helicoid spiral (Fig. 62). In many genera having the last shape, as the shell curls a solid central axis which is called the **columella** is formed. In genera without a solid columella the central space running axially up to the apex

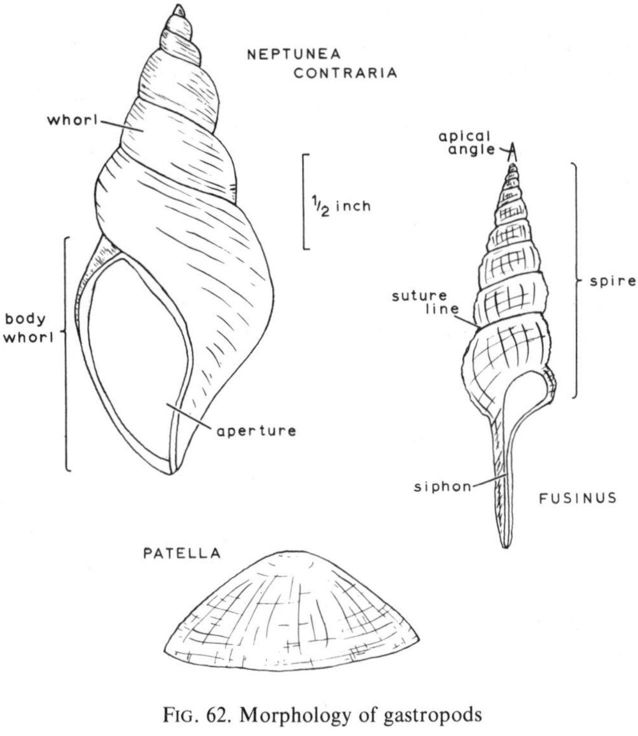

FIG. 62. Morphology of gastropods

is the **umbilicus.** Each complete revolution of the shell is a **whorl** and the last and largest into which the gastropod can retreat is the **body whorl.** Sometimes the body whorl almost completely encloses all the preceding whorls, a condition referred to as **convolute.** The **suture** or contact-line between adjacent whorls becomes less obvious under these circumstances. All the inner whorls excluding the body whorl collectively form the **spire.** In the conical gastropods the **apical angle** measured between the sloping sides of the spire is useful in classification. Most species have clockwise **(dextral)** arrangements of whorls with the **aperture** appearing on the right-hand side, e.g. *Natica* (Cretaceous to Recent) and *Littorina* (Recent). Left-handed or **sinistral** species include *Planorbis* (Jurassic to Recent) and the appropriately named *Neptunea contraria* (Pleistocene) (Fig. 62).

Apertures in general tend to be oval or crescentic-shaped and in some genera they are closed by a plate, the **operculum,** which is attached to the foot of the living creature. The operculum fits into place during the defensive withdrawal of the gastropod into its shell. Fossil shells are never found with the plates in position. Probably the most interesting structures associated with the apertures of some genera are notches around their **lip.** Occasionally the notch is short and downwards-pointing as in *Cerithium* (Tertiary) but in many forms, e.g. *Fusinus* (Tertiary) (Fig. 62), it is drawn out downwards as a narrow tube (**siphon**) three or four times longer than the aperture diameter. In life the siphon seems to act as a canal for passing fluids out from the body of the gastropod.

The external ornament of gastropod shells is very varied and consists of patterns of spiny outgrowths, ribs and knobs. They are mildly useful in classification. Modern forms are also multicoloured as no doubt fossil forms were, but colour is rapidly lost during burial and subsequent lithification.

Importance of Gastropods

Because of the difficulties in classifying gastropods, the long range of most genera, and their poor state of preservation, especially in Palaeozoic rocks, they are limited in use to geologists in solving correlation problems. Neither can they be considered certain indicators of past environmental conditions as it is not always clear which of the older species were freshwater and which marine. *Viviparus* (Jurassic to Recent), which occurs in the Purbeck beds of Dorset, is one of the few positively known to be freshwater.

Ammonites

Ammonites are the most important members, stratigraphically, of the Cephalopod group of the Mollusca. This large group which counts squids, *Nautilus*, the octopus and belemnites amongst its members is entirely marine and first became important in late Cambrian times. Since then its various members have fluctuated in numbers and genera until, at present, they seem to be on the decline. The ammonites achieved eminence during the Mesozoic era but died out suddenly and completely at the end of Cretaceous times. The reasons for this are unknown. Their beginnings were much more gradual, however, and species closely resembling them were existent in earliest Devonian times and lived until the end of the Upper Palaeozoic era. These 'proto-ammonites' are collectively called Goniatites (Fig. 63); they are used a great deal in resolving Carboniferous correlation problems.

The shells of ammonites are calcareous and most have a flat spiral shape (Fig. 63). Internally they are divided into chambers separated by partitions (**septa**). The last and largest chamber was occupied by the living creature whereas most of the others were air- or gas-filled, imparting a certain degree of buoyancy and mobility to the shell. The septa meet the outer shell along lines called **suture-lines,** which have an intricate shape consisting of numerous rounded loops. The study of the different patterns made by the loops, visible only on internal moulds and casts, is very important in classification. It is worth noting that the sutures of Goniatites are much simpler and more angular in shape.

Coiled ammonites are the dominant type and the diameter measured across the central depression (**umbilicus**) is very variable. In the Upper Jurassic limestones of Dorset giant *Titanites* are found with diameters approaching 3 ft. They are somewhat abnormal in size for Britain and most Mesozoic species have diameters of 2 to 8 in. For instance, an adult *Dactylioceras* (Lias) is about 3 in. *Dactylioceras* and *Hildoceras* (Lias) (Fig. 63) are typical **evolute** forms in which all the whorls are visible. *Schloenbachia* (Upper Cretaceous) and *Cardioceras* (Middle Jurassic) show partially **involute** or **convolute** trends, and complete involution, in which the final whorl completely encloses all the earlier, is well illustrated by *Phylloceras* (Lias). On the sides of the whorls and especially prominent on the final one in most ammonites are rib-patterns. Some ribs are straight and radial, others are sinuous and divide into smaller ribs and yet others are knobbly (**tuberculate**). As with suture-lines the rib-patterns are important in distinguishing species. The shape of the whorls in cross-section is another useful diagnostic characteristic for the palaeontologist, as great variation occurs. *Dactylioceras* has a round cross-section, *Amaltheus* (Jurassic) a narrow cone shape. In other genera it is square or of a wider cone shape. The external margin of the last whorl is called the **venter** and is commonly smooth. Occasionally it has a distinct ridge (**keel**), e.g. *Harpoceras* (Lias), or a groove (**sulcus**) as in some varieties of *Hoplites* (Cretaceous). *Hildoceras* has a combination of keel with grooves on each side. Although geologists take most interest in coiled ammonites there are a few species of Cretaceous age which are partly or completely uncoiled. *Hamites* (Fig. 63) and *Baculites* are typical representatives.

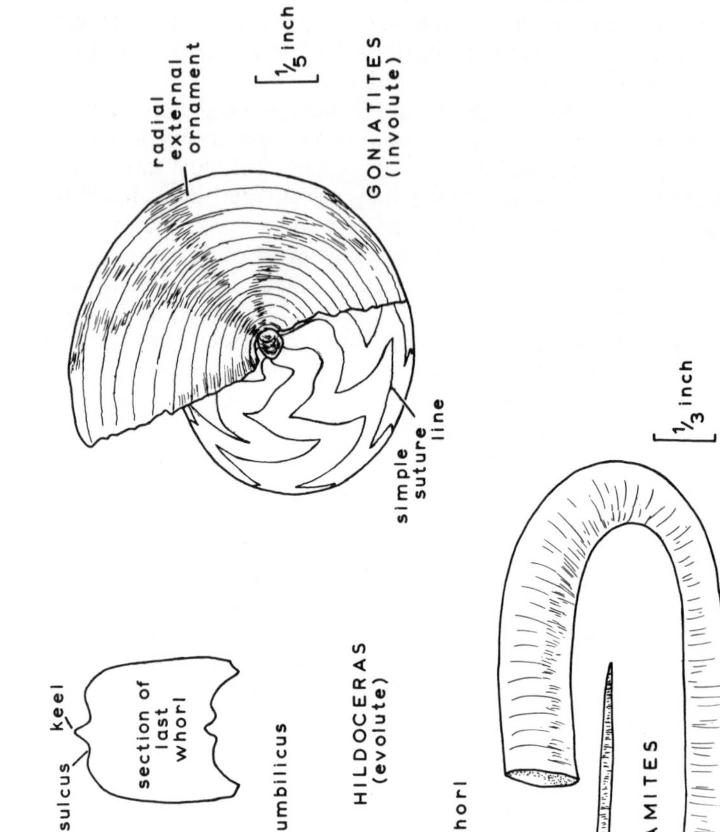

FIG. 63. Morphology of ammonites

Importance of Ammonites

The great importance of ammonites rests on the fact that they were free-floating in nature and changed their form rapidly, each form having a wide geographical spread over a short period of time. This is the perfect state of affairs for long- and short-distance correlation. The Jurassic and Cretaceous rocks of Britain have been very accurately subdivided into smaller units and accurately correlated by studying their abundant ammonite content.

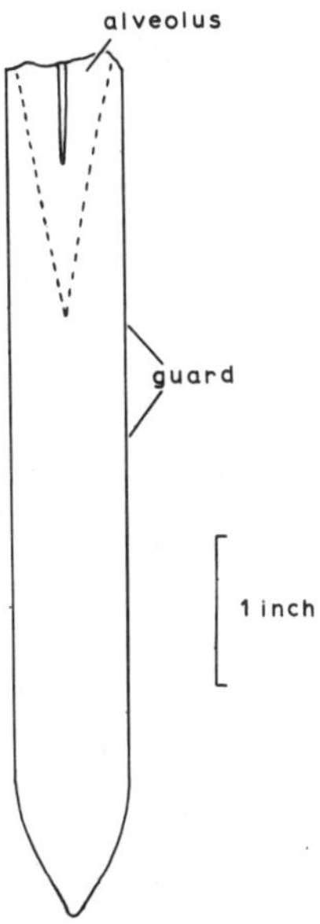

FIG. 64. Morphology of belemnites

Belemnites

Belemnites are relatively minor members of the Cephalopod group of Mollusca. They appeared in Triassic times, reached their peak in the Jurassic and Cretaceous and are now on the verge of extinction. Only one living species, *Spirula*, is known. The most important fossil genus is *Belemnites* which has a wide distribution in British Jurassic and Cretaceous marine limestones, shales and sandstones. The calcareous shells are well preserved and on the basis of their cigar-like shape (Fig. 64) it is possible to suggest that the original organism resembled a modern squid. The shells (**guards**) of belemnites may have a length of 5 to 6 in. as in *Hibolites* (Lower Cretaceous) and are usually smooth externally, apart from very narrow short grooves commencing at the top margin. If a complete guard is sliced in half lengthwise the top end can be seen to have a conical cavity (**alveolus**) of variable depth depending on the species. In life this is occupied by a chambered part of the belemnite known as the **phragmacone.** In many fossil shells only traces of the phragmacone are preserved.

Importance of Belemnites

Belemnites are free-floating marine creatures found in a variety of sediments. Unfortunately, their preserved parts do not show a great variety of structure or ornament. Hence they are not the best of fossils to use in correlating rocks unless few other organisms are present. It is partly for this reason that belemnites are used in subdividing the Lower Cretaceous Speeton Clay of Yorkshire.

Trilobites

Trilobites are marine members of the phylum Arthropoda (which includes crabs, centipedes, insects and spiders) and reached their zenith in late Cambrian and Ordovician times. They almost certainly evolved from primitive stock in the late Pre-Cambrian although there are no skeletons to prove it. By the end of Permian times they were extinct. Well over 1,000 different species have been described but very little is known in fact about the living creature. As some varieties are found in shales and others in siltstones and reef limestones it seems that they could exist in a wide range of marine environments. Some appear to have burrowed into muds for food and others probably crawled and scavenged along the bottom. A few floated and swam above the bottom and ate seaweed.

The soft parts of trilobites were protected by a flexible cover known as the **exoskeleton.** This was composed of a durable substance called chitin which is intermixed with calcium carbonate and calcium phosphate and it is the part normally preserved. Each trilobite produced many exoskeletons during life by the process of moulting. These were shed on to the sea-floor and during the intervals before a new one was secreted the animal had a spurt of growth. Most preserved exoskeletons have lengths of 3 in. or less but a few genera are exceptionally large. *Dalmanites* (Silurian to Devonian) is sometimes 14 in. long and *Terataspis* is as much as 2 ft. long. It seems odd considering the size, abundance and adaptability of trilobites that they should have died out abruptly in Permian times. Even in Carboniferous times they were becoming drastically reduced in numbers and size. Perhaps the successful incoming of 'armour-plated' ostracoderms and fish in the Devonian increased the competition for the same sorts of food. Having greater mobility the fish may soon have exerted a dominance over the trilobites in the ability to obtain food, making trilobite extinction inevitable.

The exoskeleton of a trilobite consists of three parts longitudinally, **cephalon,** flexible **thorax** and **pygidium** (Fig. 65), each

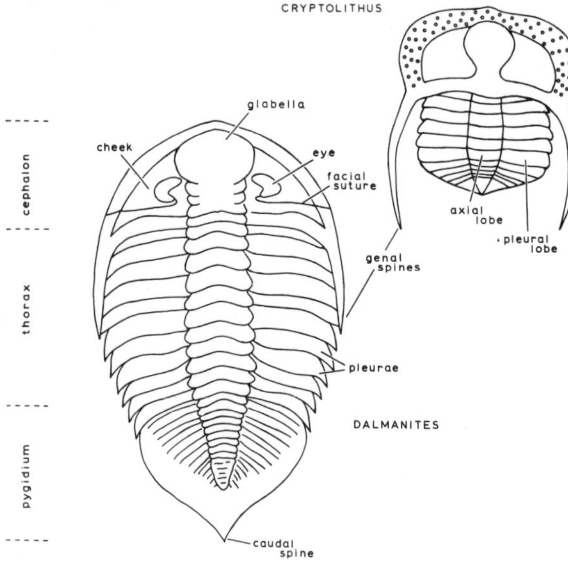

FIG. 65. Morphology of trilobites

of which may be found separately in sediments. These parts in turn are divided laterally into three. Down the middle is a rounded ridge called the **axial lobe** and on each side there are flatter **pleural lobes.** The cephalon or head shield consists of a rigid plate and is roughly semi-circular in outline. The posterior margin is straight but is sometimes prolonged at the outer edges into backwards-pointing processes known as **genal spines.** *Cryptolithus concentricus* (Ordovician) (Fig. 65) has spines which are almost twice the length of the cephalon. The central part of every head shield has a **glabella** which may be lobate, furrowed and upstanding. The glabella is oval, triangular or hemispherical in outline and occasionally extends completely across the cephalon as far as the anterior margin, e.g. *Dalmanites* (Fig. 65). At the sides of the glabella are the **cheeks** which are commonly separated into two parts by **facial sutures** which run obliquely from the anterior to the posterior margin. The sutures are lines of easy separation for the exoskeleton during moulting phases. Along the suture-lines there appear in most species **compound eyes** composed of many facets (as in modern flies). They vary in size and shape, some being large and globe-like, others being on stalks and others being hardly noticeable as in *Calymene blumenbachi* (Silurian). The kind of eye depended on the life led by the trilobite. The stalked eyes of *Dalmanites*, for instance, were suitable for bottom-grubbing in various depths of muds. The most peculiar adaptation of all, however, was that of the mud-burrowing *Cryptolithus* which was completely blind.

The thorax of trilobites consists of a series of articulated segments (up to 40 in some species) which allowed ease of locomotion and helped the organism to curl up into a neat defensive ball when necessary. Each segment has an axial lobe with **pleurae** which protected the locomotory and feeding mechanisms on either side. They are occasionally prolonged into **pleural spines** as in *Paradoxides* (Cambrian).

Most trilobites have a pygidium at their posterior end. This is a rigid plate divided into three lobes and often made up of numerous segments. It may be equal in size and shape to the cephalon, e.g. *Agnostus* (Cambrian), or grossly dissimilar, e.g. *Cryptolithus*. The pygidium of *Dalmanites* has a central spiny outgrowth or **caudal spine** at the posterior end which probably aided movement along the sea-bottom by acting as a kind of lever.

Importance of Trilobites

Trilobites reached their zenith during Cambrian and Ordovician times when most other varieties of organisms were still at an early evolutionary stage and not very abundant. Each individual trilobite also produced many exoskeletons during life and these were very durable. Consequently, their fossil remains form a relatively high proportion of the total fauna in some beds and are very valuable for solving correlation problems. The free-floating and swimming varieties are particularly important in this respect. Trilobites existed in a wide range of habitats, so they are useful in environmental reconstruction. The Lower Palaeozoic rocks of the Welsh Borderland contain abundant traces of trilobites suggestive of shelf-type deposition.

Graptolites

Graptolites are floating marine organisms of colonial habit and reached their peak development during Ordovician and early Silurian times. The more important non-dendroid forms became extinct at the end of the Silurian period, possibly owing to the predatory nature of fishes, which were just beginning to evolve in large numbers. As the dendroids became extinct in early Permian times there are no modern representatives to aid classification. Because of this the graptolites have been variously ascribed to the sponge, bryozoan and coelenterate groups of organisms. Currently it is believed that they were very primitive vertebrates capable of budding or branching during life. They are allotted, therefore, a place in the phylum Hemi-Chordata. There is little doubt that most graptolites floated passively at the mercy of tides and currents as they have a wide distribution in a variety of marine sediments. In a few cases they have been observed still attached to fragments of seaweed. Being fragile creatures with a horny chitinous skeleton they are best preserved in shales and fine-grained limestones. In most shales and slates the skeletons are commonly flattened and converted into graphite or, more rarely, pyrite. The pyritized fossils are of great value to the palaeontologist because the structural details of the original skeleton are better preserved.

The earliest graptolites appeared in late Cambrian times and, generally, remained attached to the sea-bottom in the manner of present-day seaweed. They are the dendroid kind (Fig. 66) characterized by innumerable **stipes** or branches. The stipes all branch from a small conical structure called the **sicula** which is at the **proximal** end of the graptolite. Each stipe consists of a series of cups or **thecae** in which the individual zooids grew. In most specimens their presence can be detected down one side of each stipe by a serrated or toothed margin. The shape of the external margin of thecae varies quite a lot, some being angular, others rounded or spinose. The variation is very useful in distinguishing between different species. In the dendroid *Dictyonema flabelliforme* (Cambrian) (Fig. 66) the number of stipes may be 20 or more but the more important Ordovician and Silurian non-dendroid types are characterized by 8, 4, 2 or 1. The progressive decrease in stipes seems to be a distinct evolutionary trend.

Excluding the dendroid group all the Lower Palaeozoic graptolites were free-floaters. In most species the sicula was prolonged proximally into a narrow short thread (**nema**) which could easily hook on to floating surface-debris. The stipes generally dangled downwards in a near-vertical position as in *Didymograptus murchisoni* (Ordovician) (Fig. 66) and *Monograptus* (Silurian) but in some varieties there were evolutionary modifications. *Didymograptus extensus* (Ordovician) developed two horizontal stipes and in *Tetragraptus* (Ordovician) the four stipes were turned, sometimes, through 180 degrees until they were near-vertical again.

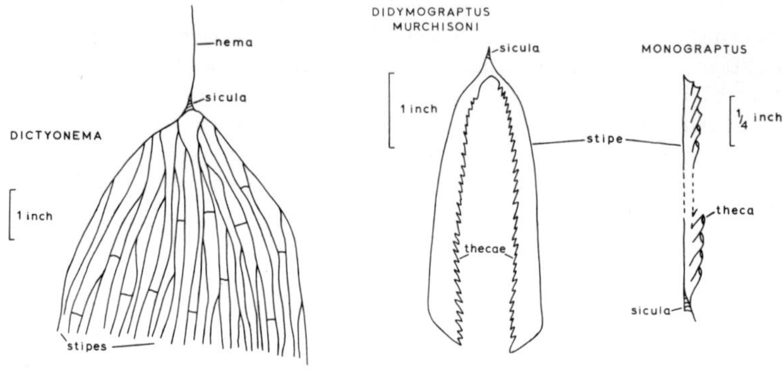

FIG. 66. Morphology of graptolites

Importance of Graptolites

Graptolites are preserved in large numbers in Lower Palaeozoic shales and slates. Most species were widely distributed at any one time. As the non-dendroid types also evolved quickly with many modifications of stipes and thecae they have proved excellent for correlation purposes. Some of the earliest geological work on the subdivision and correlation of rocks was accomplished by studying graptolites in the shales of the Southern Uplands and Wales.

Vertebrates

The vertebrates are members of the phylum Chordata and are generally defined as those animals which have backbones as a fundamental part of their skeleton. In some of the earliest fish-like vertebrates which evolved in Devonian times the backbone was very primitive but since then the bony articulated structure has become commonplace. Fish are relatively simple creatures in structure and in the organization of their nervous system. The later vertebrates, amphibians, reptiles, birds and mammals are much more advanced and modern man is the ultimate expression of complexity and organization. In these highly specialized groups the skeleton is completely protected by skin and other internal tissues.

The history of vertebrate evolution is very complicated so it will only be dealt with in briefest outline. The first fish-like creatures (ostracoderms) known in Britain occur in marine deposits of late Silurian and Devonian age. They had bodies protected by bony plates. In mid-Devonian times true scaly-skinned fishes evolved. From then onwards they spread into all manner of freshwater and marine environments. Among them were early forms of sharks and their teeth turn up frequently in marine strata (Fig. 67). A few types in the Devonian 'Age of

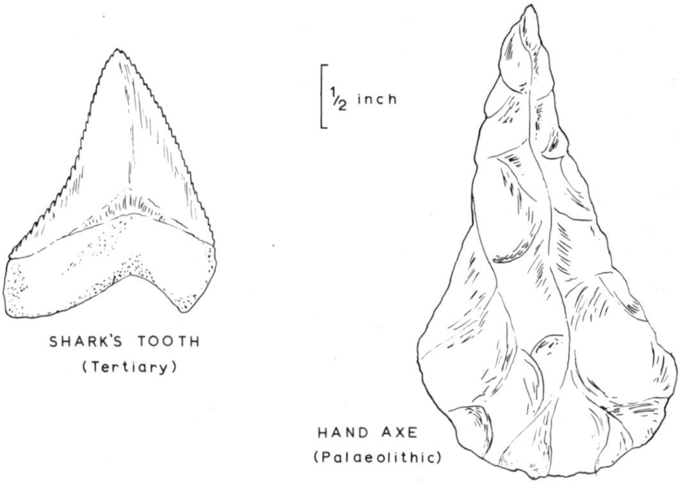

FIG. 67. Shark's tooth and hand axe

Fishes' adapted their structure so that they could slide over sea-margin muds and, eventually, they developed primitive lungs. In this way it is possible that a new major group of vertebrates known as the amphibians was created. Amphibians were capable of living for part of their life on land and part in water, and by Carboniferous times they were commonplace. Some resembled salamanders and were up to 15 ft. long, others were much smaller, a few inches in length.

The dominance of amphibians on land was soon challenged by the reptiles which no longer relied on water and so were capable of adapting themselves with much greater ease to all kinds of environments, especially arid areas. The amphibians could not compete successfully and rapidly began to decline in mid-Mesozoic times. Only a few minor varieties such as frogs and toads exist at present. The variety of reptilian life on Mesozoic land surfaces was very wide. Some reptiles were lizard-like, others like crocodiles, some were long-necked and measured more than 80 ft. from head to tail, e.g. *Diplodocus*, others such as *Brontosaurus* weighed over 20 tons. The last two are members of the dinosaur group and lived either on plants or on the flesh of other creatures. All the dinosaurs were beasts of low intelligence and were rather cumbrous in movement. Only one branch of the group, the Pterosaurs, developed any degree of mobility and this was attained by evolution of a membraneous wing. By Jurassic times *Archaeopteryx*, the ancestor of true birds capable of flapping its feathery wings as well as gliding, developed from an unknown group of reptiles.

The reptiles did not have any serious competitors until Tertiary times. Up till then the mammals, which suckle their young, were very minor members of the total vertebrate population. The first mammals seem to have evolved from reptilian ancestors in Triassic times and included forms similar to the duck-billed platypus of modern times. Some of the later Jurassic and Cretaceous varieties resembled rats and opossums. In Tertiary times the ascendancy of the mammals over all other vertebrates was completely established. All the larger reptiles and dinosaurs were now extinct. The dominance of the mammals was probably due to their greater mobility and intelligence and hence ability to seek out food and adapt themselves to changing environments. A few reptiles managed to persist, however, even up to modern times. Crocodiles, turtles, lizards, snakes and tortoises are typical specialized reptiles which were not affected by mammalian competition. The most abundant mammals in early Tertiary times were small four-legged, multi-toed creatures with primitive teeth and low intelligence. They probably ate plants and insects. Some of them almost certainly resembled miniature horses. *Eohippus* was about the size of a modern adult Alsatian dog. In later times bats first appeared together with early forms of elephants, rhinoceros, cats and dogs. Instead of chewing plants, many of these new forms had well-developed sets of teeth capable of gripping and tearing apart their prey. The teeth also began to be differentiated into different kinds, molars for grinding, canines for tearing and incisors for cutting purposes.

About the middle of the Tertiary period a group of mammals called the Primates evolved rapidly in African areas and spread outwards into Asia and Europe. They were mainly primitive apes but even in those times they had some of the characteristics of early man. They had the habit of walking and standing in a nearly erect posture and a relatively large brain constructed in a similar fashion to the human brain. In contrast to man they had abnormally elongated arms, fang-like canine teeth, big toes capable of grasping branches and insufficient intelligence to think constructively, communicate their ideas or make implements suitable for mastering their environment. It is very probable that both these apes and man evolved from some common ancestor at about this time. Possibly the ancestor was a near relative of *Australopithecus*, Southern Ape, which had more man-like features than most apes. The skulls of *Australopithecus* were found in 1925 and 1936 in a series of South African limestone deposits. These deposits are difficult to date accurately but are probably late Pliocene or early Pleistocene, that is approximately 1 to 5 million years old. By middle and late Pleistocene times man as a social and organized being of superior intellect to other mammals was established in most parts of the world. Geographical isolation prevented mixing between the various centres of evolution and, consequently, many distinct types of man developed. The skull of Swanscombe Man found in north Kent in association with hand axes (Fig. 67) of palaeolithic type differed in many ways from the Java or Peking Man who lived at approximately the same time but many thousands of miles away. Differences in the facial characteristics of races can still be seen at present despite the fact that interbreeding is commonplace.

Besides the burst of Primate evolution culminating in man, the Tertiary period also saw the development of most of the modern vertebrates. Flying and wading birds became widespread in nearly all land and sea-margin environments early on and modern-looking fish, mainly bony, began to abound in rivers, lakes and seas simultaneously. Elephants and carnivorous (flesh-eating) beasts such as sabre-toothed tigers and bears increased prolifically for a short time and spread into many continents. Later on they gradually tended to become extinct or much more restricted geographically and in numbers. Elephants, for instance, are confined now to southern Asia and Africa and are slowly decreasing in number. In part this decrease is due to the destructive activities of man.

Plants

The earliest plant life known occurs fossilized in some Pre-Cambrian marine limestones of Rhodesia and North America which are 2,000 million years old. The fossil traces consist of laminated and fibrous calcareous nodules up to 2 ft. in diameter and bear a resemblance to modern algae or seaweed. In some instances these old algae have accumulated on the sea-floor to form distinctive reef-bodies as in the Great Belt Series limestones of Montana. Many reef-algae have a fine net-like structure and encrusting habit which acts so as to bind loose debris together into a rigid whole. It is a quality retained by some algae even at the present day and modern coral-reefs have a high algal content. In modern cold-water lakes and seas there are also primitive algae called diatoms. They have a relatively simple siliceous skeleton and are free-floating near to the surface, where they appear to form the staple diet of some fishes. The skeletons are so prolific, however, that death creates diatomaceous oozes over wide areas on the sea-bottom. In contrast to calcareous algae they have no capacity to build reefs and none have been found in rocks older than the Cretaceous.

Towards the end of Silurian times and certainly during the Devonian the evolution of the first land-plants took place. At first they were very primitive and small. Proper root systems had not quite developed neither were there proper leaves. Instead, the stems and branches were either completely smooth or spiny although capable of producing spores in large numbers. Some of these early plants are preserved in the Devonian Rhynie chert, Scotland—a petrified peat. By Carboniferous times in the northern hemisphere the vegetation on land was profuse, particularly in hot, steamy swamp-areas. Some evergreen trees such as *Lepidodendron* (Fig. 68) and *Sigillaria* grew to heights of 80 ft. or more. *Calamites*, horsetails with hollow stems, attained heights of 50 ft. and grew densely in marshy groves. Their modern equivalents still prefer damp soils but rarely reach a size of more than 4 ft. Ferns also evolved during the Carboniferous period.

During the Mesozoic era ferns became even more widespread as did conifers and cycads. They all remained dominant until late Cretaceous times when flowering plants (angiosperms) appeared for the first time. These and deciduous trees such as oaks, planes and poplars gradually gained ascendancy in the Tertiary period. Simultaneously many varieties of grass began to extend over land-surfaces where drier climates prevailed. The innumerable varieties of plants by now were almost as varied as in modern times and capable of adapting themselves to changing environmental conditions. For example, the plant life in north-western Europe during Tertiary times was predominantly a warm-climate type but this quickly altered under the influence of colder air and ice-sheets spreading from the north. Tropical palm trees and subtropical evergreens migrated to warmer African areas and were replaced by a scantier cold-climate flora. Since the Pleistocene ice-age this northern flora has progressively become enriched with varieties compatible with the present milder and wetter climate.

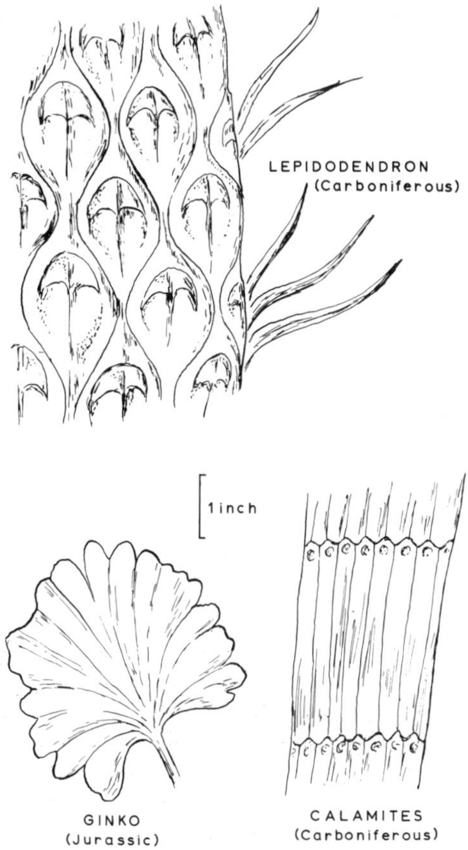

FIG. 68. Plants. Only a relatively small part of each plant is represented

Importance of Fossils

The preceding descriptions of the main animal and plant groups have indicated their value to the geologist. Above all else they have shown that it is possible to reconstruct the distribution of land and sea at any given time in the past. In doing so it is clear that the cumulative evidence of all the fossils present in a series of beds is much more satisfactory than the study of one group.

Even so the geologist, in reconstructing palaeogeographies from fossil traces, has to be very cautious because many organisms may not be in their position of growth. Local currents in seas or river-erosion on land may move the skeletons many miles and deposit them in an environment totally different from that of life. A more extreme example of this occurs when fossils are worn out of older lithified beds and incorporated into younger sediments, as in Surrey where the Cretaceous Bargate Beds contain fossils derived from older Jurassic limestones. Accurate palaeogeographic pictures can only be built up, therefore, if the geologist takes into account the various sedimentation factors in operation at the time and relates them to the fossil distribution. There are many organisms, of course, which remain in position of life unaffected by currents. Reefs formed of corals and algae are prime examples. Old reefs are also excellent for the study of the direct relationship of corals, algae, and countless other reef-loving creatures to each other and to their immediate environment. This kind of study is called **palaeoecology.**

In working out the palaeoecology of fossil communities it sometimes becomes evident that the assemblage also indicates palaeoclimatic conditions. Plant debris in the Carboniferous rocks of Spitzbergen in the Arctic Ocean and in Permian rocks of Antarctica suggest hotter climates than at present. Coral-reefs in the Silurian, Carboniferous and Jurassic rocks of Britain also suggest much warmer climatic conditions at those times.

One of the most intriguing aspects of fossils is the way they show evolutionary trends over a given period of time. The stipes of graptolites progressively decreased in number through the Lower Palaeozoic era. The simple suture-lines of some Palaeozoic cephalopods became much more complex in Mesozoic times and the typical coiling of ammonites became less pronounced in the late Cretaceous. These are just a few detailed examples of the broad process of organic evolution, a process which has been continuous since at least late Pre-Cambrian times. Many groups of organisms commenced their evolutionary cycle in a very minor way then became widespread in a 'burst of evolution' and finally faded away to negligible numbers, e.g. the reptiles. In contrast, graptolites and rugose corals developed and became extinct much more rapidly for no clearly defined reason. Then there are Lingula-type brachiopods which have hardly altered at all from Palaeozoic up to modern times and seem to have always existed in a specific kind of muddy environment. They are good examples of what is called, rather oddly, 'static evolution'. The most important evidence of progressive evolution, however, is that of man's descent from primitive mammalian ancestors in the last 10 million years or so.

The importance of any fossil, whether used to infer palaeo-environments or deduce evolutionary processes, depends on the firm knowledge of its position in time and space. For more than a hundred and fifty years it has been known that strata superimposed on each other contain different assemblages of fossils. Because of these differences it is possible to subdivide strata into groups each containing a distinctive fauna or flora. In many cases the organisms are so distinctive that rocks containing them can be equated from one region to another. This process is called correlation. In the earliest days of systematic geological study strata could only be correlated in a general way. Subsequently fossil assemblages have been studied in such refined detail that very thin subdivisions can be traced over many thousands of square miles. The smallest subdivision is known as a **zone** and each has within it an assemblage of organisms from which is usually selected a representative **index fossil.** The best zonal and index fossils are goniatites, ammonites, graptolites and some varieties of foraminifera because they had a floating habit, wide geographical spread and modified their shape and ornament frequently. Corals, sponges, algae and brachiopods are less good because of their adherence to the bottom and their need for specialized growth-conditions.

QUESTIONS

1 Draw, label and name one variety of coral, trilobite and graptolite and describe the kind of environment in which each lived.

2 Select two groups of organisms which are good for zonal purposes and two groups which are not and explain, with illustrations, the reasons for your choice.

Chapter 7

ROCK CHRONOLOGY

THE subdivision and time-correlation of the rocks of the earth's crust have always been an exciting challenge to geologists since William Smith's discovery in the early nineteenth century that the rocks in England and Wales varied in fossil-content from place to place. Smith's ideas were soon extended by more detailed studies of fossils throughout Britain and the Continent and by the turn of the century all fossiliferous rocks had been divided into major time-units or **systems,** each characterized by distinctive assemblages of organisms and, frequently, by equally distinctive rock assemblages. Apart from time differences further studies showed that assemblages of the same age varied when traced laterally. For instance, marine limestone-shale deposits sometimes pass into non-marine sandstones and conglomerates when followed at outcrop. This effect gave rise to a new term **facies,** meaning the total faunal and lithological characteristics of a group of rocks which distinguish it from its time-equivalents elsewhere. Simultaneously the relationships of the beds were worked out using the **Law of Superposition,** which states that providing inversion has not occurred, the uppermost beds in any exposed section are the youngest. The many groups of rocks could then be arranged in chronological order or in the form of a **stratigraphical (geological) column** with the oldest at the base and the youngest at the top (Fig. 69).

In unfossiliferous strata correlation is very difficult, especially if exposures are poor and few in number. Even with excellent exposures time-correlation is likely to be inaccurate over distances of more than 20 miles. This is because beds which are identical at outcrop may not have been deposited at exactly the same time. Uniform beds are capable of cutting across time-planes as is well illustrated by some Liassic sandstones which can be traced from Gloucestershire into Somerset and then southwards into Dorset. In Gloucestershire where they are called the Cotswold Sands they were deposited earlier than in Somerset (Yeovil Sands) or in Dorset (Bridport Sands) (Fig. 70).

The age of sediments in the stratigraphical column can never be determined absolutely from the fossils they contain or from the types and thicknesses of beds. This is because little is known about the rate at which fossils and sediments accumulate on the sea-bottom. If the rate was known, then a simple sum of multiplying the total thickness of a group of rocks by the time it takes for 1 in. to accumulate would give the accurate age of those rocks. Unfortunately sedimentation is very irregular so that any age assessment using it must be very approximate. In present

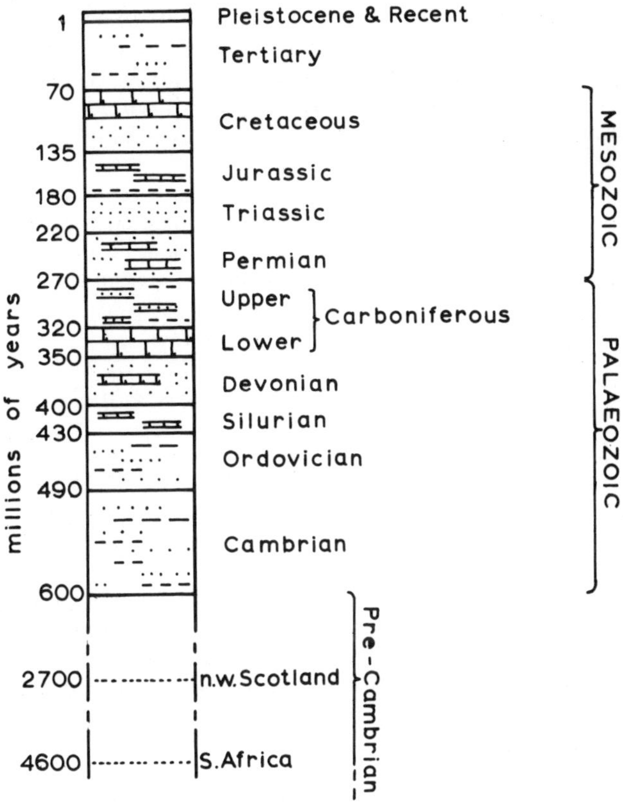

FIG. 69. Stratigraphical column

oceans and seas deposition averages about 6 in. in 1,000 years. By applying this information to older rocks and relying on the **Law of Uniformitarianism** which says that general conditions in the past were no different from those of today it is possible to arrive at a rough dating for various lengths of time such as the Mesozoic era or Devonian period. The Lower Carboniferous rocks in Britain were estimated by these means to have been deposited about 250 million years ago, the Cretaceous Chalk about 55 million years ago, and so on. These time-values are generally satisfactory for most geological purposes but it was soon realized that the absolute dating of rocks would be even more satisfactory, especially for the great length of Pre-Cambrian time. A few geologists in the 1930's began working on radio-active minerals such as pitchblende, biotite and glauconite hoping that they would provide the answer. The basis of their work rested on the fact that the rate of breakdown of a radio-active mineral into a related inert mineral could be calculated. By determining the proportion of old radio-active to newly-created inert matter in any of these minerals the age of the containing rock would then be known with greater accuracy than by any other method. The ideas of the early workers were enthusiastically received and the last 30 years or so has been a period of extensive research on a wide range of radio-active minerals. The most interesting result of this method of dating has been the lengthening of time-periods

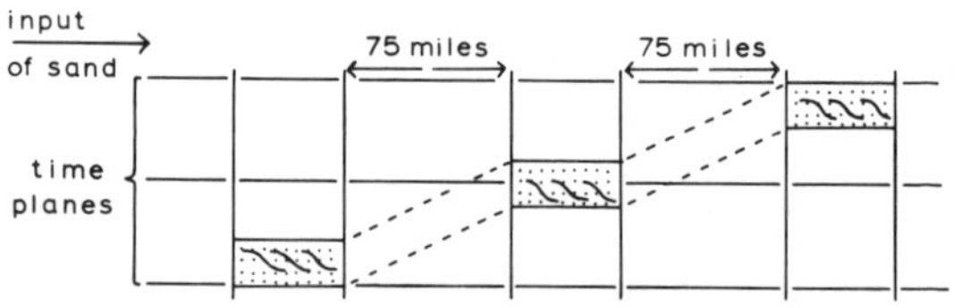

FIG. 70. Transgression of time-planes. The same sandstone becomes progressively younger the further it is away from the source

in the stratigraphical column. The Lower Carboniferous rocks, for example, were laid down some 330 million years ago (not 250 million) and the Chalk was deposited about 80 million years ago (not 55 million). The oldest rocks of the earth's crust are in South Africa and are about 4,600 million years old. They are much older than the Pre-Cambrian rocks of north-west Scotland which are between 2,500 and 2,900 million years old.

With the help of organisms, rocks and modern radio-active dating the stratigraphy, that is the correlation and palaeogeography, of many areas can be inferred now with reasonable accuracy. Only the British Isles area will be discussed here.

Pre-Cambrian Times

It is impossible to build up a cohesive history of the Pre-Cambrian rocks of Britain because the outcrops are widely spaced and unfossiliferous, apart from some algal structures found at Charnwood. Additionally, insufficient radio-active determinations have been made so far to correlate the individual outcrops accurately.

The most extensive outcrop occurs in the Northern Highlands of Scotland where a thick series of partly red-stained, cross-bedded and slumped sandstones, interbedded with grey and red shales, rest with marked angular unconformity on a highly metamorphosed basement-complex of metaquartzites, schists and gneisses. Equally complexly folded metamorphic rocks also occur in Anglesey, north Wales and the Malvern Hills, but whether they are identical in age to the Scottish ones is problematical. In Pembrokeshire, Shropshire and the Charnwood Forest there are ashes, agglomerates, lavas and sediments, whereas in the northern Pennines and Shropshire there are thick successions of conglomerates, greywackes and slates. It is usually assumed that the sediments are younger Pre-Cambrian than the metamorphics and this is demonstrably true in Scotland. Elsewhere in Britain there is little proof for this assumption.

Lower Palaeozoic Times

The Lower Palaeozoic era in the British Isles area commenced about 600 million years ago with a widespread marine transgression across a very irregular land-surface formed of Pre-Cambrian rocks. Consequently, wherever the basal contact between the two groups of rocks is seen there is a marked unconformity. In north Wales, the Malvern Hills, Shropshire around the Wrekin and Caer Caradoc, and in the north-west Highlands the basal beds of Cambrian age are rather unfossiliferous beach-conglomerates and orthoquartzose sandstones. They pass upwards into shales yielding trilobites (*Paradoxides*, *Olenellus*, *Agnostus*) and brachiopods (*Lingulella*). Immediately to the north of the Midland Valley of Scotland the junction is much more obscure and cannot be pin-pointed at any one locality or by any one particular level in the rocks. One reason for this is the folding and metamorphism of the rocks. The faunal difficulties in defining the base of the Cambrian system here are related in part to the nature of the sedimentation during those times, in part to the paucity of fossils. There is little doubt that the conglomerates and well-sorted sandstones of north-west Scotland and the Welsh borderland were deposited adjacent to land-masses. Ripple marks, cross-bedding and worm-borings are just a few of the typical shallow-water features which confirm this. Away from the land and its marginal shelf-sedimentation areas the character of the deposition was quite different. Central Wales, central Ireland, the Lake District and the Southern Uplands and Central Highlands of Scotland were the sites of basins or downsags in the earth's crust. They are referred to collectively as the Lower

Palaeozoic geosyncline and within them sediments accumulated very quickly and thickly under relatively deep-water conditions. Conglomerates, greywacke sandstones with graded bedding, shales and occasional manganese ores were the main deposits during Cambrian times and they are poorly fossiliferous.

The general distribution of land and sea and the various facies within the sea remained similar during Ordovician times apart

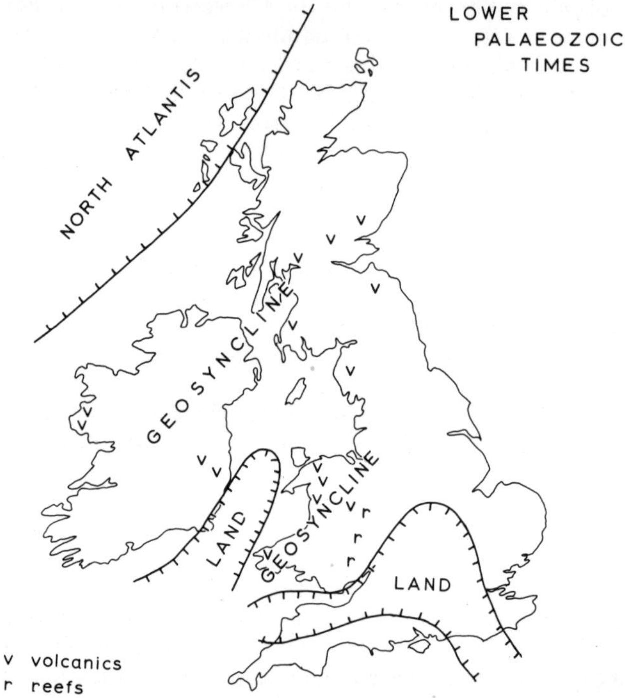

FIG. 71. Lower Palaeozoic palaeogeography

from the Midland Valley area of Scotland, where the waters shallowed sufficiently for a rich shelly fauna to be created and shelf-like sediments deposited. No land was in the immediate vicinity of these shoals. The axial regions of the geosyncline seem to have been particularly unstable at this time with vulcanicity and earth-tremors commonplace. Great thicknesses of rhyolite, andesite and basalt lavas and tuffs erupted in the Lake District, north Wales and the Southern Uplands. Simultaneously there were abundant intrusions such as dykes, sills and laccoliths formed of granophyre, dolerite and gabbro and stocks of serpentinite. The present-day rugged topography of the Snowdonia and Lake District mountains is due largely to the weathering of these hard resistant igneous rocks.

As the axial areas of the geosyncline were very unstable it is not surprising that the shelf facies reflects this instability. The main expression was not so much in the form of igneous activity but in slight tilting of the shelves with consequent minor changes in the position of the shore-line. The seas advanced and retreated many times across the shelves and in doing so created a number of relatively minor unconformities.

Faunally the Ordovician period is very interesting because graptolites, e.g. *Didymograptus*, were evolving rapidly and spreading into all sea-areas. Trilobites, e.g. *Ogygiocaris*, were also much more numerous and adapted themselves to most muddy, silty and calcareous environments. Generally they preferred shallow waters, as did a whole host of brachiopods (*Orthis*), tabulate corals, gastropods, lamellibranchs and cephalopods (*Orthoceras*). The shelf-sediments are thus very fossiliferous and because of this are sometimes collectively called the 'shelly facies'. In contrast, the deeper water sediments are usually poorly fossiliferous, apart from graptolite-rich shaly layers. The term 'graptolite facies' is sometimes substituted for geosynclinal facies because of this, despite the fact that graptolites are found in the shelf facies as well.

In Silurian times the broad picture of sedimentation was maintained with rapidly deposited and occasionally slumped greywackes and shales in the axial regions and thinner deposition of calcareous shales, limestones and sandstones on the unstable marginal shelf-seas. At this period in central Wales much of the clastic debris was being swept in by currents from the south and south-west and the land-areas to the east were contributing little. The eastern shelf-areas were sometimes occupied, therefore, by clear marine waters in which reefs and calcareous organisms could proliferate. The Wenlock Limestone of the Welsh Borderland consists in part of reefs which extended in barrier-fashion along a rough north–south line for more than 30 miles. *Halysites*, *Favosites*, stromatoporoids and bryozoans were the reef-builders. Other clear-water limey phases were ideal for the accumulation of thick shell-banks of *Conchidium*, *Chonetes* and rhynconellids.

Caledonian Earth-movements

By the end of Silurian times something like a 7 mile thickness of sediment had accumulated in the axial parts of the Lower

Palaeozoic geosyncline. Instability of the whole region by now was becoming even more pronounced and shallow water, red-stained sands and shales began to spread over both axial and shelf-areas. The change in conditions was such that the graptolites died out completely and the scanty fauna in the rocks consisted mainly of primitive fish, ostracoderms and brachiopods such as *Lingula*. Clearly something more than minor oscillations of the earth's crust was taking place and, in fact, these widespread changes in sediments and organisms marked the beginnings of the Caledonian earth-movements which continued into Middle Old Red Sandstone times. Areas which had previously been shallow and deep-sea began to rise upwards to form extensive tracts of land and this land immediately came under the attack of subaerial weathering agents. The main effects of compression and uplift were in the axial regions of the geosyncline. By Middle O.R.S. times the excellent roofing-slates of north Wales had been created and the rocks highly folded and faulted. The greywacke–slate belt of the Southern Uplands was another area of particularly intense compression and the beds were tightly folded along north-east to south-west lines so that today there is much repetition of outcrop. Large batholithic intrusions of granite became emplaced at a late stage during the movements in the Lake District, Southern Uplands, Scottish Highlands and Ireland. Perhaps the most spectacular evidence of profound earth-movements at this time is the widespread thermo-dynamic metamorphism of rocks north of the Midland Valley of Scotland. There the forces must have been very powerful—so powerful that not only were many of the rocks completely reconstituted but also pushed bodily to the north-west along north-east to south-west trending thrust-fault planes. Some of these deep-seated thrust planes, e.g. the Moine thrust, are now exposed at the surface in the Assynt area of north-west Scotland.

Devonian Times

Throughout most of Devonian times the presence of the Caledonian movements had an important influence on sedimentation. Around the flanks of the hills and mountains uplifted in central Wales, Ireland and Scotland thick and irregular deposition of fluviatile conglomerates (well seen near Edinburgh), sands and muds were laid down. Their main field-characteristic is a red colour caused by haematite which was precipitated under the oxidizing conditions prevalent at those times. Hence the appropriate name of Old Red Sandstone which is used for this facies of the Devonian. Within the estuaries, distributaries and inland freshwater lakes ostracoderms and fish were sometimes prolific, particularly in the Orcadian Province of north-east Scotland. Occasionally they seem to have been killed off rapidly, possibly by changes in salinity; their remains may be seen covering some bedding planes. There is little doubt that the lakes dried out often because evaporite seams and chemically precipitated calcite layers and nodules are intermingled with red sun-cracked

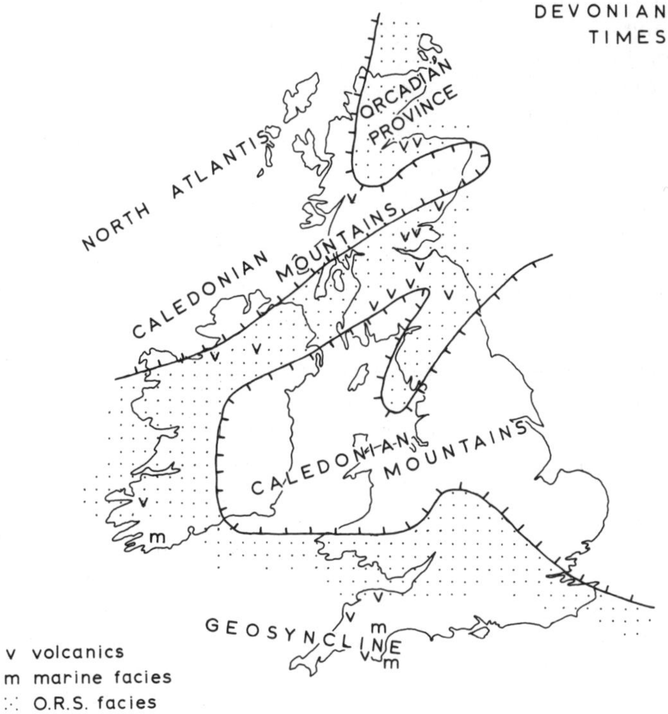

FIG. 72. Devonian palaeogeography

and rain-pitted muds, marls and silts. The nodular calcite bands are called cornstones.

While intermontane deposition was taking place in northern Ireland and Scotland and marginal deposition in southern Ireland and south Wales, the crustal instability of Lower O.R.S. times was commonly expressed by subaerial volcanic eruptions and the outpouring of thick piles of andesite and trachyte lavas. In the Midland Valley the Sidlaw, Ochil and Pentland Hills with their characteristic stepped or 'trap' profile all have a foundation of these lavas. Lavas were also erupted in Cornwall and Devon but there they were basaltic in nature and submarine. The most important feature of the submarine lavas, however, is their lateral passage into marine limestones and shales. In the Torquay district they are interbedded with limestones, which contain reef-building corals such as *Pachypora* and *Cyathophyllum* and stromatoporoids (encrusting organisms), marine shales and sandstones with *Spirifer* and early varieties of goniatites. It is obvious, therefore, that in the South-West Peninsula area the conditions of deposition were predominantly marine and quite different from the terrestrial O.R.S. facies to the north. The marine Devonian was laid down in shallow waters in an east–west trending basin extending across the Channel and into southernmost Ireland. Northwards a progressive transition into fluviatile red conglomerates and sandstones occurred. The transition is particularly well seen in north Devon where three northwards-thickening wedges of red sandstone are interbedded with Devonian marine shales. This is important for correlation between the two facies which would otherwise be almost impossible.

Lower Carboniferous Times (Carboniferous Limestone)

The Carboniferous period began with an important marine transgression. The fluviatile O.R.S. flats began to be inundated by the sea which extended across Cornwall and Devon. The pattern of sedimentation in Cornwall and Devon also changed as more sandy material was deposited than in Devonian times. The sandy material appears to have been derived from a newly-created land-mass to the south called Cornubia which occupied part of the western English Channel. This thick succession of sands and silts interbedded with shales, cherts and coal-films is called the Culm Measures.

The effects of the marine transgression were felt first on the southern side of St. George's Land, a residual land-mass in the position of the Midlands. Very quickly the seas moved around the western and probably the eastern end of St. George's Land and extended over most of Ireland, northern England and the eastern end of the Midland Valley basin. Most of the sediments were laid down on shallow-water shelves. As St. George's Land was very low-lying and supplying little clastic debris the waters adjacent to it were sufficiently clear for many varieties of limestone and occasional anhydrite beds to be formed. In south Wales and the Bristol area thick bryozoan, crinoidal, algal, oolitic and dolomitic limestones were deposited. In many of them there is a rich fauna of corals such as *Dibunophyllum*, *Lithostrotion* and

Caninia and brachiopods such as *Productus* and *Schizophoria.* In central Ireland, the Isle of Man, Derbyshire and Malham districts reef limestones, some algal-rich, were also laid down.

The land-mass to the north of Britain called North Atlantis was a much more hilly area than St. George's Land and continued to supply clastic debris, so that marine sediments became less dominant on approaching it. Fluviatile sandstones and shales

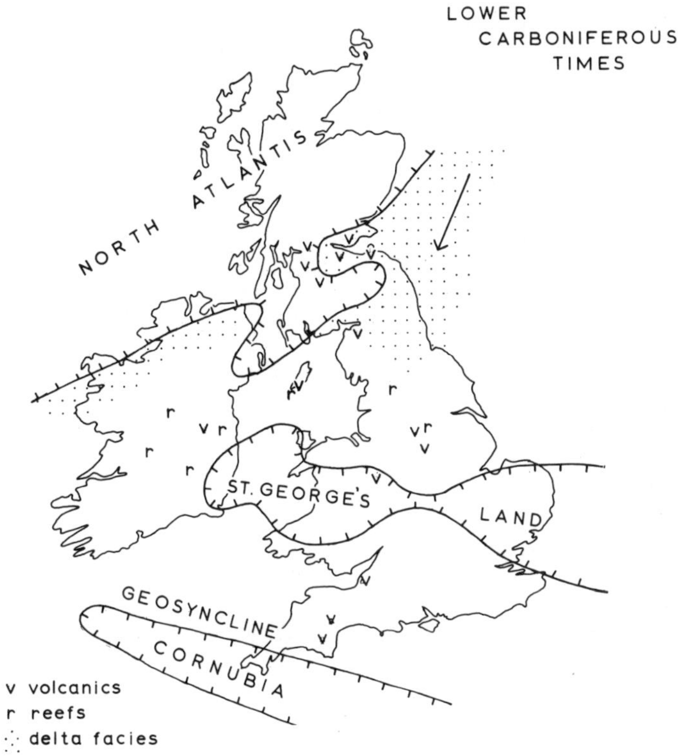

FIG. 73. Lower Carboniferous palaeogeography

were spread around it by major rivers and these periodically inundated the shallow marine seas. The most important delta occupied the site of the North Sea and its western distributaries poured abundant material into the eastern side of the Midland Valley basin and into northern England. On some of the sand- and mud-banks vegetation flourished occasionally and eventually it was converted into coals and seat-earths, while in some of the quieter lagoons bituminous shales and oil-shales were created, particularly near Edinburgh. The maximum extent of sandy deltaic sedimentation at this time was as far south as the Lake District and Yorkshire Dales. The final tonguing-out of the delta might be represented by the black shales with goniatites, solitary corals (*Zaphrentis*) and lamellibranchs (*Posidonia*) in Derbyshire.

Vulcanicity was rife in all areas and localized earth-tremors must have been commonplace. In Derbyshire and around the

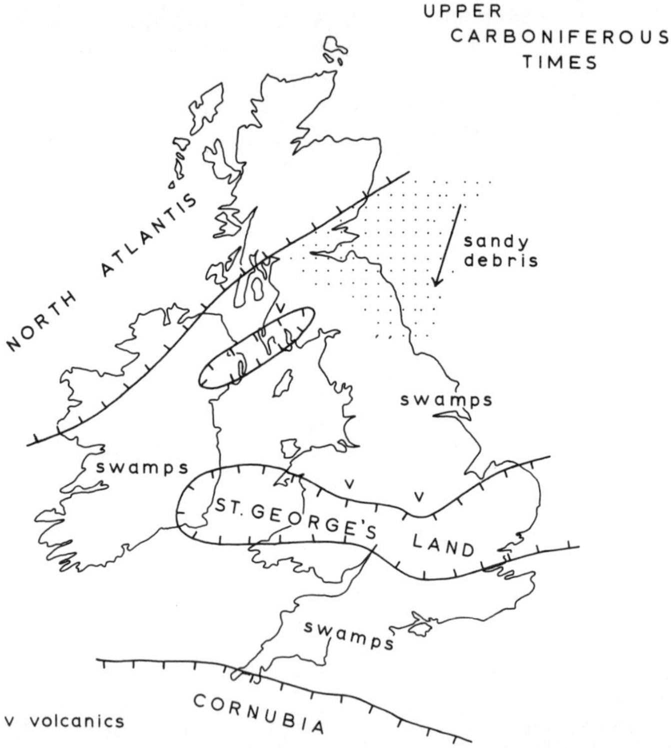

FIG. 74. Upper Carboniferous palaeogeography

Firth of Forth the numerous volcanoes probably started life under a shallow cover of water and some basaltic lava-flows were almost certainly submarine. The larger eruptive centres quickly built up above water-level, however, to form subaerial cones covering many square miles. One of the largest probably extended over most of the city of Edinburgh, although only residual traces now remain in the prominent features of Arthur's Seat and the basalt plug on which the castle stands. The greatest outpouring of subaerial basalt lavas was at the western end of the Midland Valley where more than 2,000 ft. accumulated very quickly. The tops of each lava weathered into red lateritic soil indicating a subtropical climate at the time and the recent weathering of the hard lavas and soft laterite layers gives the typical 'trap' topography as seen in the Campsie Fells and Clyde Plateau.

Upper Carboniferous Times (Millstone Grit and Coal Measures)

At the end of the Lower Carboniferous there occurred a phase of earth-movement which uplifted North Atlantis, St. George's Land and Cornubia. These lands immediately succumbed to rapid erosion and the material was deposited in adjacent sea-areas. This early phase of rapid sedimentation is called Millstone Grit times. The dominant influence to the north of St. George's Land continued to be the major north to south delta system but by now its western distributaries were very powerful and carried much more fluviatile debris than before. Hence southern Scotland and northern England were quickly inundated by coarse sands and muds on which vegetation often flourished. During any prolonged and relatively quiet muddy phases marine organisms re-colonized the broad delta flats so that some of the shales contain goniatites and lamellibranchs such as *Pterinopecten* and *Posidonia.* When the fluviatile influence was too great for true marine creatures, more tolerant inarticulate brachiopods such as *Lingula* and *Orbiculoidea* persisted, together with non-marine lamellibranchs.

To the south and west of St. George's Land sedimentation was similar although the fluviatile sands and muds appear to have been laid down by a series of relatively minor rivers issuing from St. George's Land and Cornubia.

By later Carboniferous or Coal Measure times the various lands supplying debris were doing so at a slower rate. Simultaneously the basins of deposition were becoming choked with debris and stagnant swamp-like conditions were very widespread. Within these swamps through which distributaries now flowed sluggishly conditions were ideal for a very varied fauna and flora. The more important were the plants, e.g. giant horsetails, *Lepidodendron*, *Sigillaria*, which grew densely in the humid tropical environment and eventually died, rotted and formed coal-seams. Sometimes the simple pattern of organic growth and clastic deposition was disturbed by crustal warping and marine waters from the south and east of the British Isles invaded the swamps, killing off the plants for a short time. These marine episodes terminated a non-marine phase of deposition and as they were repeated or partly repeated several times the rock succession is said to show a cyclicity or rhythm in deposition (Fig. 75). The initial surge of fresh fluviatile material after a marine episode was occasionally so powerful that the underlying rocks were eroded so as to create shallow, wide meandering channels. These are wash-outs or, as the miner calls them, 'wants'. Seams they are working are suddenly cut off by the sandstone or shale infilling the channels.

The coal-miners also find that seams are destroyed by igneous intrusions. The Midland Valley was particularly prone to intrusion by dolerite and gabbro sills and dykes while Coal Measure deposition was still taking place. Stirling Castle is built on a prominent crag of dolerite which is a sill as is the gabbroic Salisbury Craigs which dominate the southern part of Edinburgh.

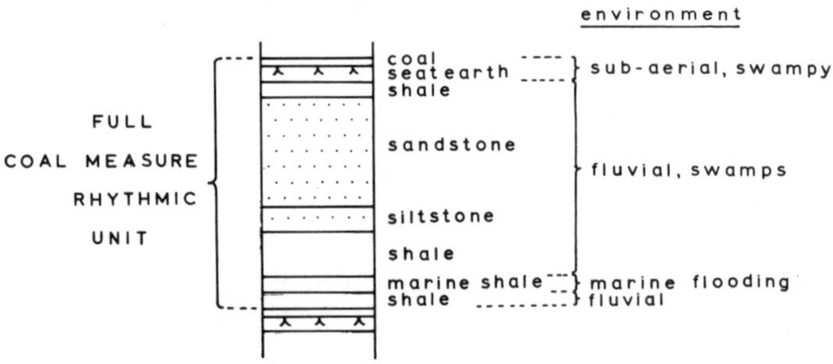

FIG. 75. Coal Measures rhythmic unit

Elsewhere in Britain igneous activity was uncommon apart from in Leicestershire, where basalt lavas and sills were erupted for a considerable period starting in Millstone Grit times.

The end of the Carboniferous was marked by slight earth-movements and a return to a more continental style of deposition with rather unfossiliferous red, haematite-rich conglomerates, sandstones and shales. A few non-marine lamellibranchs (*Anthraconauta*) and fish persisted in shallow muddy lagoons and plants were still common in some damper areas but, generally, organic life was negligible compared with earlier Carboniferous times. These conditions were the earliest indication that Britain was about to pass through a phase of comparative aridity.

Armorican (Hercynian) Earth-movements

The earliest phase of these movements, named after the old term for Brittany, Armorica, and the Harz Mountains of Germany, occurred in late Carboniferous times. The major phase, however, involving the upheaval of all the pre-existing rocks, their folding, faulting and erosion was at the end of the Carboniferous period. The depositional trough between Cornubia and St. George's Land bore the brunt of the compressional forces and was subject to intensive folding and thrust-faulting along east–west lines. In the Ammanford area of the south Wales coalfield the coals were squeezed so much that they changed into high-ranking anthracites. Cornwall and Devon had the additional complication of granite intrusion followed by copper, tin and iron mineralization. At a late stage in the batholithic emplacement of the granite magma, as at Dartmoor and St. Austell, hot magmatic fluids altered the cooling granite into china-clay, greisen and schorl rock.

To the north of St. George's Land block-faulting and broad open-folding were the main structural effects. The Lake District and most of Scotland and Ireland were uplifted slowly. The old red soil profiles created on this new land during the pre-Permian interval are still preserved in many places. Intrusions of basic igneous rocks were still common, one of the best known being the dolerite Whin Sill of the northern Pennines. At about the same time extensive lead-zinc mineralization took place, mainly affecting the Carboniferous Limestone areas of Flintshire, the Derbyshire Dome and northern Pennines.

Permo-Triassic Times

The general aridity of the depositional basins heralded in the late Carboniferous reached its peak soon after the Armorican movements. The newly-created highlands surrounding the basins protected the latter from excessive rainfall and created low-lying desert conditions. From the highland areas large quantities of pebbles, sands and muds and red iron oxides were supplied by spasmodic torrential floods and these spread into the basins in fan-like fashion. The finest material was then partially redistributed by winds. The red coloration shown by most of the sediments is so widespread that the name New Red Sandstone is used by many geologists. The Brockrams of the Vale of Eden are reddened breccio-conglomerates derived from the erosion of the Lake District hills. The breccio-conglomerates surrounding what appear to have been islands of Carboniferous Limestone in south Wales and the Mendips are also haematite-stained. Sometimes the wind-transported materials accumulated into cross-bedded dunes but on other occasions they were deposited in short-lived playa lakes in which gypsum was being precipitated intermittently. In both instances the sand grains show good rounding and frosted surfaces. A few lakes in the Cheshire Basin were relatively large and unusually persistent and within them evaporation reached such great proportions that rock-salt was precipitated. The greatest precipitation of salt minerals took place, however, in an arm of the great Zechstein Sea which extended from Russia and northern Germany into northern England as far west as the Lake District. At first the sea was normal

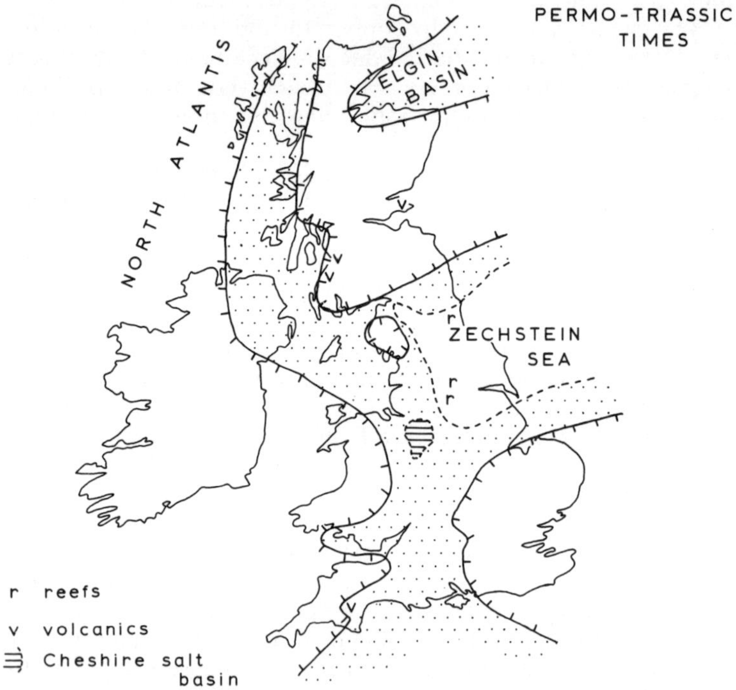

FIG. 76. Permo-Triassic palaeogeography

open marine with abundant Productids and algal and shelly reefs living and forming in a clear calcareous environment. These limestones, now dolomitized, are well exposed from Durham into Nottinghamshire. Later the sea contracted in size owing to extreme evaporation and thick layers of rock salt, gypsum and anhydrite were precipitated. These are not very well exposed at the surface and have mainly been detected by deep boreholes put down north-west of Whitby through a thick cover of younger Jurassic rocks.

Because of the hot semi-arid circumstances vegetation was

sparse on the land-areas but this did not deter the spread of large amphibians and small reptiles into most environments. A few reptiles were beaked and probably lived on plants but most were spiky-toothed flesh-eaters. Primitive bony fishes and lung-fishes occasionally colonized the lakes and seas but very rarely abounded.

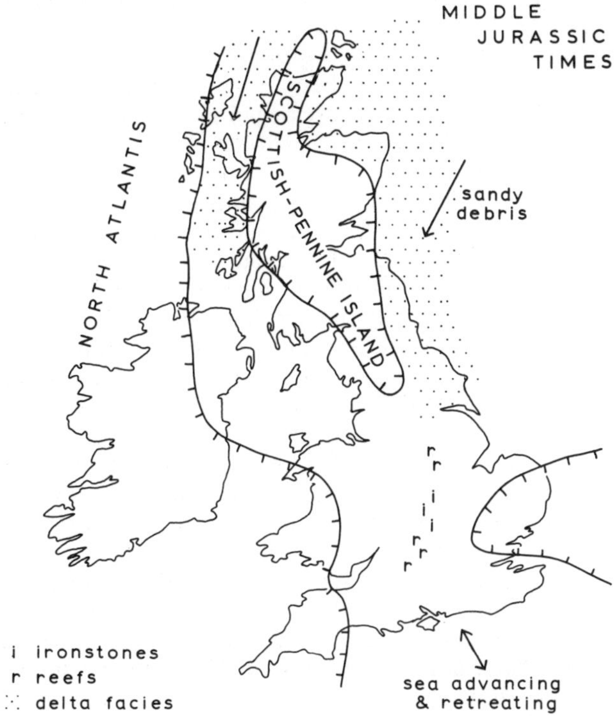

FIG. 77. Middle Jurassic palaeogeography

Jurassic Times

During the Jurassic period the climate remained hot. As the mountainous belts to the north, west and east of the British Isles were now much lower-lying the rainfall was more evenly spread over land and sea. Very little coarse debris was being derived from the land and this, combined with continued down-warping of the land-surface, seems to have allowed the seas to transgress from the south and east. In the earliest Jurassic times, therefore, thin marine muds and limestones were laid down over most of the South and Midlands and as far north as the Inner Hebrides and Moray Firth on each side of the Scottish-Pennine island. Only rarely was this uniform pattern of sedimentation disturbed. In the Cotswolds and Dorset area deltaic sands were deposited and in Lincolnshire and Yorkshire oolitic ironstones of some economic value. Within all these rocks, collectively referred to as the Lias, organic life was prolific. Ammonites and marine reptiles of the Ichthyosaur and Plesiosaur type were commonplace.

In the middle Jurassic (Fig. 77) minor earth-movements began to affect the distribution of sediments. From the north a large input of fluviatile sand and clay began to extend southwards in a very similar way to that of the Millstone Grit deltas. On the deltaic sand- and mud-flats in east Yorkshire thick vegetation grew. This eventually formed poor quality coal-seams. The plants were quite different from those of the Carboniferous period and there were many varieties of cycads and conifers. The thick sands and muds thin out towards the Midlands where the northern deltaic influence was only mildly felt. Here, shallow marine waters generally prevailed, in which coral-reefs proliferated and oolitic and bioclastic limestones were deposited in association with marine shales. Oolitic ironstones of great economic value were formed in Northamptonshire. They give every indication of shallow waters with cross-bedding and cut-and-fill structures, abraded ooliths and comminuted shell debris. Southwards into the Cotswolds the facies again changes relatively quickly into massive oolitic limestones and marine shales with an abundant fauna including corals (*Isastraea*, *Thecosmilia*), lamellibranchs (*Gryphaea*, *Ostrea*), brachiopods (*Terebratula*, *Rhynchonella*), echinoids (*Clypeus*, *Cidaris*) and crinoids (*Apiocrinus*). Marine influences were clearly dominant in this area. The massive limestones now form the impressive westward-facing escarpments of the Cotswold Hills. They have been quarried extensively for building-stone, particularly near Bath.

Towards the end of the Jurassic period the marine seas again transgressed northwards into Scotland and widespread thick muds and limestones accumulated. The Oxford and Kimmeridge Clays, rich in ammonites such as *Kosmoceras* and reptiles, are the best-known, the former having firing qualities suitable for the production of bricks. The transgressive phase soon terminated as a result of earth-movements which caused the seas to retreat eastwards into Germany and southwards into France. Most of north Britain was uplifted and in East Anglia and the east Midlands the seas shallowed so drastically that a land-barrier known as the London Platform began to emerge. This barrier in due course almost separated northern and southern Britain into two distinct areas of sedimentation. To the south of it clear waters prevailed and thick marine limestones with *Aptyxiella*, the 'Portland Screw', and giant ammonites were deposited together with marls and sandstones. The limestones include the Portland Stone, which is amongst the best building-stones in the world and is mainly quarried on the Isle of Portland. Soon even these conditions were replaced as freshwater streams and brackish lagoons extended southwards. Within them thin limestones and ostracod-rich shales were formed. Occasionally the waters shallowed sufficiently for forests of cycads and conifers to grow out of good soils. Residual traces of both can be seen at Swanage, Lulworth Cove and the Isle of Portland. The thickest development of these non-marine Purbeckian rocks is exposed on the coast to the south of Swanage and there the limestones ('marbles') and gypsiferous marls and shales contain abundant gastropods (*Paludina*) and lamellibranchs (*Unio*) intermingled with fish, mammal and turtle debris. At a few levels in the succession there are thin black 'cinder-beds' composed of marine lamellibranchs and these suggest that true marine waters were never too far distant and rapidly advanced northwards if the supply of terrestrial debris ceased for any reason.

North of the London Platform uplift was so pronounced that all the earlier rocks, including the Kimmeridge Clay, were now land and being eroded. Only in Lincolnshire did sandy marine deposition continue without any great break and this suggests the persistence of a residual tongue of the eastern sea.

Cretaceous Times

In the earliest part of the Cretaceous period the distribution of land and sea remained more or less as it was during the late Jurassic (Fig. 78). Earth-movements were still taking place and northern Britain began to downwarp slowly towards the east, allowing the eastern sea to transgress once more into Yorkshire and Lincolnshire. In Yorkshire thick marine muds rich in belemnites, ammonites and fish debris were laid down. These are the Speeton Clays and they rest with unconformity on all the older rocks. South of the Humber estuary sedimentation seems to have been more or less continuous through from Jurassic times and was markedly different from that of Yorkshire. Instead of homogeneous clays a mixed series of sandstones, marls, oolitic ironstones, limestones and pyritic clays were deposited.

In contrast to the downwarp of the northern land towards the North Sea basin, the land-areas of the London Platform, Wales and south-west England on which vegetation was growing profusely, e.g. cycads, conifers, ferns and flowers, seem to have been sharply uplifted at this time. Rapid erosion and deltaic deposition began immediately to the south and west of the London Platform and the non-marine sediments of late Jurassic times were totally and conformably inundated by the fluvial materials of the Wealden Beds. The pattern of sedimentation established

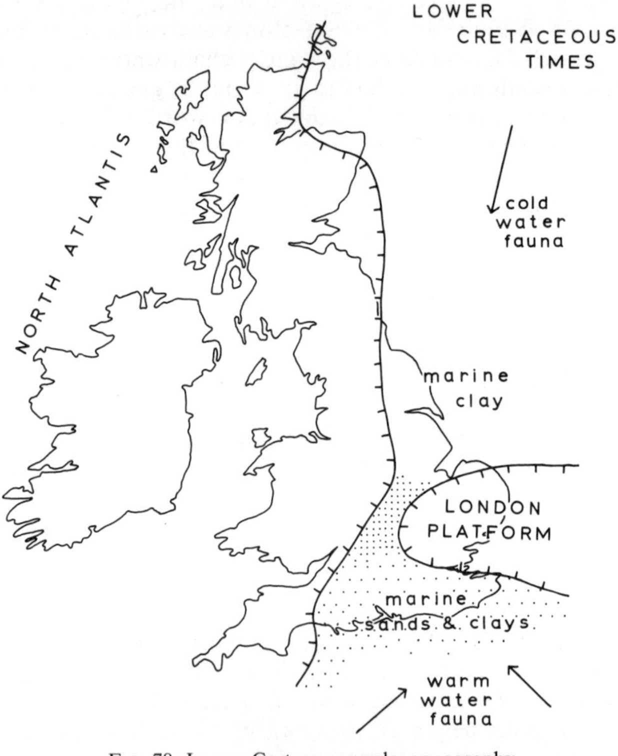

FIG. 78. Lower Cretaceous palaeogeography

was very similar to that of the Upper Carboniferous and mid-Jurassic deltas with cross-bedded sandstones and clays interbedded with thinner ironstone seams, fossil soils and uncoalified layers of vegetation (*Equisetites*). At their maximum extent the deltas probably spread well into northern France. Towards their source-areas and especially at the western end of the London Platform the beds thin out very rapidly.

Towards the middle of Cretaceous times slight earth-movements, combined with a slackening in the supply of debris from adjacent land-areas, allowed warm marine seas rich in brachiopods, sponges, bryozoa and ammonites (*Hoplites*) to invade southern Britain once again. Muds and green glauconitic sands were deposited, hence the name Lower Greensand given to the group of rocks. At outcrop the sands are normally so highly weathered that much of the glauconite and other green clay minerals have broken down into yellow-brown limonite. The sea quickly spread round the western end of the London Platform and a positive link with the northern colder-water sea was firmly re-established for the first time since the deposition of the Jurassic Kimmeridge Clay. The sediments are relatively thin and mainly sandy around the western and northern flanks of the Platform but pass into clays as they are traced away into Yorkshire.

Yet another phase of earth-movement followed the Lower Greensand transgression and this time the Platform and adjacent areas were completely submerged and covered by clays and greensands. Within the Gault Clay, which is especially thick in the Weald, the coiled ammonite *Hoplites* is found as well as the uncoiled *Hamites*. Belemnites, lamellibranchs and brachiopods are also common. As the Gault is traced westwards it gradually changes into the Upper Greensand with an abundant marine fauna. The most peculiar facies-change of these times, however, was that near to the old northern margin of the London Platform in Norfolk. There the sand-clay facies of the south passed into red chalk and this chalk can be traced into Yorkshire. Clearly the Platform was still influencing deposition even though now completely inundated. The red chalk was the precursor of the almost complete submergence of the British Isles by the Chalk Sea during late Cretaceous times. By the end of this Chalk Sea transgression only residual low-lying land-areas persisted in northern Scotland and, possibly, southern Ireland. These contributed little clastic debris apart from shore-line sands into the Chalk Sea which was warm, shallow and clear. Microfossils such as algae abounded and on death accumulated by the million to form the organic calcareous ooze later to be known as the Chalk. Large organisms were less tolerant of the conditions so that they were very sparsely distributed. They include echinoids (*Micraster*, *Holaster*), lamellibranchs (*Inoceramus*), brachiopods (*Terebratulina*) and ammonites (*Schloenbachia*). The Chalk (and its characteristic layers of flints) is well exposed in the Downs and Chiltern Hills where it forms a good scarp-and-dip feature. In Lincolnshire the Wolds are less pronounced topographically but the Chalk quickly regains its eminence in eastern Yorkshire, especially along the southern flanks of the Vale of Pickering. Along most of its outcrop it is worked for lime and in some areas, e.g. northern Kent, it is mixed with local clays and river-muds to produce cement.

Tertiary Times

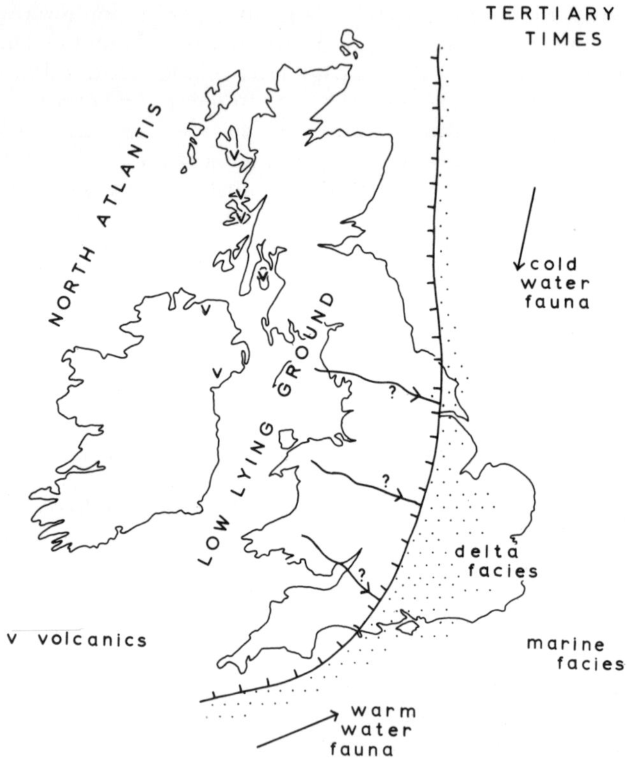

FIG. 79. Tertiary palaeogeography

The British Isles as a whole warped slightly at the end of the Cretaceous period and the subsequently deposited Tertiary marine clays, limestones and fluviatile sands rest with gentle unconformity on the older rocks. In the Isle of Purbeck sandy Tertiary gravels can be seen resting on a deeply pot-holed Chalk surface even though the beds are dipping quite concordantly. Land now existed everywhere in Britain except East Anglia, the London Basin and Hampshire Basin and on it plant life was very prolific with tropical palms and evergreens. Mammals of many sizes and shapes roamed through the forests and grasslands, and crocodiles and turtles abounded in shallow inland lakes and rivers. The rivers disgorging from the land were powerful, consequently deltaic-type sedimentation was common in the adjacent seas. Poor quality lignite seams and fireclays sometimes formed on top of the cross-bedded sands and much plant debris was incorporated in the muds. Thick pebble-beds, e.g. the Blackheath Beds, composed almost entirely of flints derived from the Chalk, were also deposited over large areas. The great deltaic spreads extended far into Belgium and northern France at their maximum extent. Frequently, however, the deltaic deposition slackened off and marine waters reinvaded the area bringing with them a marine fauna similar in character to that found at present in warmer seas. Marine gastropods abound in some of the clays and include *Murex*, *Cerithium* and *Fusinus*. The London Clay contains *Nautilus*, *Turritella* and the echinoid *Hemiaster* intermingled with many other marine organisms and plant debris. Marine facies deposition sometimes extended as far west as the Isle of Wight and Dorset.

Whilst the cyclical deposition was taking place in England, northern Ireland and western Scotland were being disturbed by powerful earth-tremors and extensive vulcanicity. Large volcanoes centred on the Mourne Mountains, Arran, Mull, Skye and Ardnamurchan erupted continuously and from them there poured out countless basalt lava-flows which spread out over Antrim, the Inner Hebrides and the western end of the Midland Valley. Most of the thick pile of lavas were subaerial and developed lateritic weathered tops but some flowed into shallow lakes where they became interbedded with plant-rich shales and coals. Linear dolerite dyke swarms were intruded simultaneously along roughly north-west to south-east lines and a few individual dykes even managed to extend into northern Yorkshire and the Midlands. Some dykes may have acted as feeders for lava-flows at this time.

In middle to late Tertiary times the gentle earth-movements which had been affecting the British Isles became more intense. This powerful phase is known as the Alpine earth-movement because the Alps were upheaved at the time. South-eastern Britain was uplifted, folded and faulted along east–west lines and the sea retreated south-westwards towards the Bay of Biscay and eastwards towards the North Sea. Deltaic and estuarine deposits formed upon the margins of the North Sea basin at this time are now found in East Anglia and are commonly referred to as ‘crags’. They are very shelly with an early dominance of warm-water organisms followed, in later crags, by a progressive incoming of colder-water Arctic-type organisms.

The Alpine fold-movements which brought about the retreat of the seas were very important because they broadly created the present shape of the British Isles. They were still connected, however, to the Continent by a land-bridge on the site of the English Channel. The Wealden anticlinorium was formed together with the highly asymmetrical folds of the Isle of Wight and Weymouth areas. On updomed surfaces complicated drainage patterns slowly evolved which today give rise to the characteristic topography of most of south-eastern England. Other features of importance connected with the Alpine movements were the haematite replacements of the Carboniferous Limestone on the western side of the Lake District and the lead mineralization of the Mendips.

Pleistocene and Recent Times

The colder-water fauna in the later crag-deposits marked the oncoming of widespread cold conditions which reached their peak in Pleistocene times. The glacial sculpting of the British Isles has already been mentioned (p. 101 and Fig. 51) so need not be discussed in detail. It is only sufficient to note that the ice-sheets advanced and retreated several times though never further south than the Thames–Bristol Channel line. During the warmer interglacial phases conditions were such that elephants, hippopotamuses, hyenas and lions roamed freely. Primitive man colonized riverside sites as at Swanscombe and by using flint implements managed to eke out a precarious existence.

The final retreat of the ice-sheets about 10,000 years ago was marked by oscillations of the land and sea due to isostatic and eustatic adjustments. The numerous raised beaches around the British coasts date from this period as do the ria coastlines and incised gorges of inland areas. The most important change was the severing of the last land connection with the Continent and the creation of the Straits of Dover. By this time the flora and fauna of the British Isles was almost identical with that found today.

Sea-floor Spreading, Plates and Continental Drift

To piece together the geological history of the British Isles it is necessary to draw upon all the main branches of geology. The work is difficult but, as more and more precise factual evidence of all kinds accumulates so the palaeogeographical picture becomes clearer and more realistically portrayed. The basic geological principles used in the reconstruction of old geographies are the same whichever part of the world is being considered.

Within the last few years certain remarkable facts have been discovered which are very important in unifying the story of the Earth's history. The ideas they confirm are far from new but, until now, these ideas have had to rest on weak foundations due to lack of data. The first information comes from in-place basalts occurring along some of the deep oceanic ridges, such as the Mid-Atlantic Ridge (Fig. 25). The basalts are younger in the volcanically active axial parts of the ridges and become progressively older towards the volcanically inert margins. In other words, the Earth's crust is being added to in the axial parts and simultaneously is being forced apart sideways. This effect is known as **sea-floor spreading** and may amount to 2–3 inches per year (Fig. 80).

While new outer crust is being added so is pre-existing older crust being carried down into deeper crustal layers. These areas of downward movement are very noticeable around the margins of the Pacific Ocean, where they are marked by elongated deep sea basins and trenches roughly paralleling the adjacent land areas. They can also be areas of active mountain building and deep seated earth tremors, because the pushing down of crustal material commonly leads to a buckling and snapping of any overlying sedimentary layers. The passage of crust into regions of higher pressure and temperature gives partial remelting of the solid rock, so that intrusive batholithic masses of granite and ultrabasic rock are created.

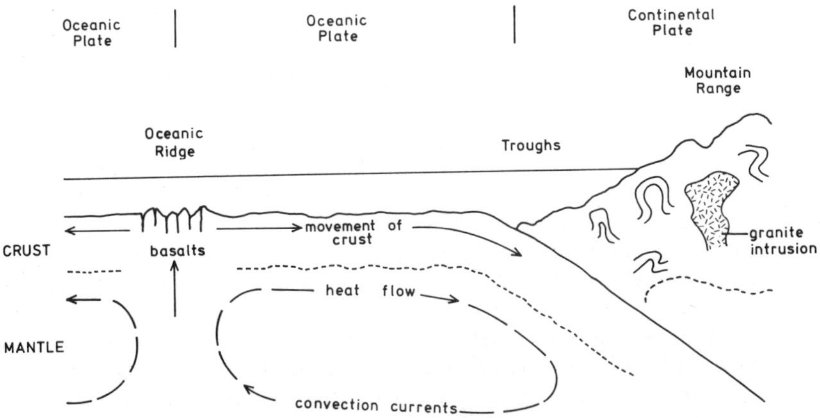

FIG. 80. Convection, plates and sea-floor spreading. The convection currents probably result from heat generated during the breakdown of radio-active minerals. The zone of compression, fracture and melting between oceanic and continental plates is usually steeply dipping

So, the situation is that the top 60–100 miles of the Earth's crust is now known to be capable of movement, with crust being continually added in some places and subtracted in others. The relatively narrow zones where these additive and subtractive processes mainly occur are now believed to mark the edges of relatively rigid parts of the crust known as **plates.** Convection currents within the mantle (Fig. 26) probably play an important role in the location of these edges. There are about 6 or so large plates, e.g.'s Pacific (an oceanic type), America (a continental type) and a few small plates constituting the surface of the Earth. The larger ones are several thousands of miles across.

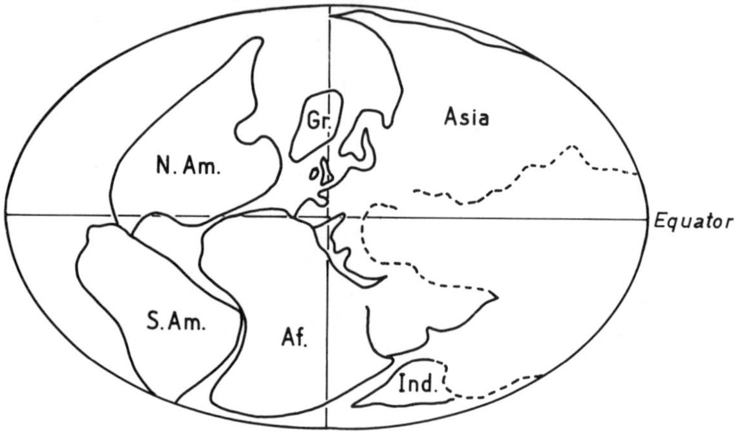

FIG. 81. Continental masses [plates] during Carboniferous times. Note the way in which the masses fitted together before they drifted apart in subsequent eras

As plates seem to be able to move in a conveyor-like fashion over the mantle there is a very important consequence. Sediment layers resting on the plates, whether thin as on ocean bottoms or thick as on continents, are displaced sideways. Whole continents may be slowly moved to occupy positions many hundreds or thousands of miles from where they were first established. This is known as **continental drift.** Figure 81 illustrates the probable position of the major continental plates about 275 million years ago (Carboniferous times). In Lower Jurassic times (about 170 million years ago) Australia and Antarctica seem to have separated from Africa and South America. Faunal evidence as well as other geological facts suggest that the South American and African plates began to drift apart in Lower Cretaceous times (about

120 million years ago), so opening up the South Atlantic Ocean. This process continued through into Upper Cretaceous–Tertiary time leading to the complete opening of the North Atlantic and Arctic Oceans and the separation of Australia from Antarctica. Generally, the present arrangement of continents and oceans was reached by late-Tertiary times.

QUESTIONS

1 Describe the various methods of subdividing geological time.
2 Either (*a*) describe and account for the variation in deposition during Devonian times, or (*b*) compare and contrast Lower Carboniferous sedimentation with that of Middle Jurassic times.
3 Briefly describe the effects of the Caledonian and Alpine earth-movements in Britain.
4 Describe three of the following: (*a*) the Lower Palaeozoic geosynclinal facies, (*b*) Upper Carboniferous and Middle Jurassic deltas, (*c*) Permo-Triassic salt basins, (*d*) the Chalk Sea, (*e*) Tertiary vulcanicity.

Appendix 1

FIELD WORK

GEOLOGY is a field science which means that any serious study must begin outdoors by examining the rocks at outcrop. This process is known as establishing the field relationships of the rocks and by understanding them problems which appear very difficult in the laboratory may be quickly resolved. A basalt hand specimen, for example, cannot indicate whether it was intrusive or extrusive at its time of formation but the field relationships can.

The equipment needed by a geologist to carry out field work is basically very simple. All he requires is a compass for recording directions of dips, trends of fold axes and any other features of structural interest, a clinometer for measuring the amount of dip, a notebook, and a base map, contoured if possible, on which to insert relevant information. It is usual to carry a hammer and hand lens ($\times$ 10) for a closer inspection of rocks at outcrop. The main function of the hammer is for the collection of specimens for laboratory work. Many diagnostic features of rocks are much better seen if weathered surfaces, not fresh, are closely inspected. Oolitic texture and fossil traces are commonly much more obvious in weathered rocks.

The collection of representative specimens of minerals, rocks and fossils in any given area is very important, both from the point of view of working out their geological history and for future reference in case the outcrops are built over or covered up by tipping. Many classic localities noted for particular assemblages of rocks and fossils have been lost in this way. The field specimens have to be transported back to the laboratory so it is essential that they should be wrapped up carefully and given some reference number. Reference numbers, which should be duplicated in a notebook, are extremely important because the exact locality and bed from which the specimen has been collected should be known. As soon as the field location becomes unknown then the specimen loses most of its geological value.

The locality should also be recorded on a map together with other relevant data such as the dip of the beds, their type, and so on. The prime function of a geological map, however, is to show the outcrops of beds. As soon as specific data are recorded the geologist tries to link them up on his map to form a cohesive pattern. In doing so every available natural feature is used.

Where there is a thin top-soil a hand auger is very helpful for sampling the underlying solid rock. Normally 6 in. to 1 mile topographical maps are used as a base for the construction of geological maps but if more detailed information has to be inserted then the 25 in. scale is adopted. In unexplored ground and where a rapid preliminary survey is required the maps used may have a much smaller scale of 1 in. to 1 mile and detail has to be omitted. Mapping in new areas is usually made even more complicated by the absence of any topographical base maps in which case the geologist has to construct his own. It is a time-consuming process but can be speeded up by studying aerial photographs and transposing topographical information from them on to the uncontoured base map.

In mapping new ground geologists are handicapped by a lack of preliminary information whereas if they are remapping old ground there is usually a fair amount of data already collected. In the British Isles the latter situation is normal because most areas have been mapped by the Geological Survey over the last 100 years though in variable detail. Many of the Geological Survey maps are available to the general public at a low cost. Commonly associated with each 1 in. map there is a detailed summary, or District Memoir, of the geology of the area with many references to specific localities. Regional Guides are also published but they lack detail simply because they cover much larger areas. There are, therefore, ample sources of official information for most British outcrops and these can be supplemented by reference to a large body of information within numerous geological society publications.

In collecting specimens and mapping, geologists are fortunate because many exposures in road-cuttings, quarries and cliff-faces are freely accessible. There are, however, numerous occasions when exposures occur in privately-owned ground. Permission to enter that ground must be sought and, usually, is freely given. At the same time great care should be exercised in observing the Country Code of Conduct and not leaving rock fragments strewn about the routes of access and exit as these may cause serious damage to farm animals.

Appendix 2

LABORATORY WORK

LABORATORY studies follow naturally from field work. The minerals, rocks and fossils collected have to be identified correctly, labelled as to their age and locality, and carefully stored. In many instances it is usual to photograph or draw the actual specimens so that their subsequent loss is not too serious a matter. Fossils are the most important group in this respect, especially as many tend to be fragile. Some fossils are so small that drawings need to be enlarged in order to bring out the important morphological details whereas others may be so large that reduction is necessary. It is also advisable to draw whole and solid fossils from two or better still three viewpoints so as to indicate their shape in three dimensions.

Rocks and minerals are much more difficult to deal with because their features may not be capable of adequate illustration. There is little to be gained by drawing fine-grained mudstones, limestones, hornfelses or basalts unless they show, in addition, obvious structural and textural features such as lamination, cross-lamination, porphyroblasts, phenocrysts and vesicles. Likewise with massive, irregular mineral specimens. Minerals need drawing only if they show a good habit.

Whether capable of illustration or not all rocks and minerals can be described. In rocks the important features are mineral content, colour, grain size, evenness of texture and fossil content (if any). Minerals have a wide range of physical properties (see Chapter 2) which can be described, although some properties are more diagnostic than others and need emphasizing, e.g. the red streak of haematite.

The proper identification of fossils, rocks and minerals has a direct relevance to map-making and geologists normally correct field observations, if necessary, at this stage. The throw of faults, pitch of folds and other structural complications are deduced as well. The experience gained during the working out of problem maps using strike lines comes in very useful at this point and forms the basis of accurate completion of the map. It is essential that a sound grounding in solving problem maps should have been gained at some early stage. Their importance is twofold because they not only help to solve field problems but also help to build up mentally a three-dimensional picture of the rocks. Block-diagrams are also very useful in this respect.

Appendix 3

FURTHER READING

BRADSHAW, M. J. and JARMAN, E. A., 1969, *Reading Geological Maps*: English University Press.

British Regional Geology guides: H.M. Stationery Office.

DAVIES, A. M., 1961, *An Introduction to Palaeontology*: Murby.

GILLULY, J., WATERS, A. C. and WOODFORD, A. O., 1959, *Principles of Geology*: Freeman.

GREENSMITH, J. T., 1967, *Practical Geology for Schools*: Leonard Hill.

HILLS, E. S., 1953, *Outlines of Structural Geology*: Methuen.

KIRKALDY, J. F., 1967, *Fossils in Colour*: Blandford.

KIRKALDY, J. F., 1968, *Rocks and Minerals in Colour*: Blandford.

KIRKALDY, J. F., 1971, *General Principles of Geology*: Hutchinson.

OAKLEY, K. P. and MUIR-WOOD, H. M., 1956, *The Succession of Life through Geological Time*: Brit. Mus. (N.H.).

READ, H. H., 1971, *Rutley's Elements of Minerology*: Murby.

WELLS, A. K. and KIRKALDY, J. F., 1968, *Outlines of Historical Geology*: Murby.

INDEX